Praise for One of the Bestselling, Most Influential, and Most Widely Talked About Business and Management Books Today

"ONE OF THE TOP TEN BUSINESS BOOKS OF ALL TIME."
  —Xerox Business Services Magazine

"MEG WHEATLEY GAVE THE WORLD A NEW WAY OF THINKING ABOUT ORGANIZATIONS with her revolutionary application of the natural sciences to business management. . . . Her ideas have found welcome homes in the military, not-for-profit organizations, public schools, health care and churches as well as in corporations. . . ."
  —American Society for Training and Development

"A BOOK LIKE *LEADERSHIP AND THE NEW SCIENCE* ONLY COMES ALONG ONCE IN A DECADE. Margaret Wheatley pushes our thinking about people and organizations to a new dimension. You will never think about organizational life in the same way again."
  —Ken Blanchard, coauthor of *The One Minute Manager*

"IF THERE'S A SINGLE BOOK THAT SETS THE STAGE FOR THE FUTURE OF ORGANIZATIONS, THIS IS IT . . . Wheatley makes complex ideas simple, and then shows how those simple ideas can be used as powerful tools."
  —Stephen E. Ewing, President and CEO, Michigan Consolidated Gas Company

"*LEADERSHIP AND THE NEW SCIENCE* SURPASSES ALL BOOKS TO DATE IN MANAGEMENT SCIENCE. It is truly in a class by itself, introducing a standard of excellence in thought and perception against which all other management books and thought will surely be measured."
  —Gerene Schmidt, Founder and CEO, Science, Business, and Education, Inc.

"IF YOU WANT TO THINK ABOUT CHANGE AND ORGANIZATIONS IN AN ENTIRELY NEW WAY . . . *READ THIS BOOK NOW.*"
  —John R. Berry, Vice President, Corporate Communications, Herman Miller, Inc.

"HOLD ONTO THE TOP OF YOUR HEAD WHEN YOU READ THIS BOOK. . . . Using exciting breakthroughs in biology, chemistry, and especially quantum physics, Wheatley paints a brand-new picture of business management. this new relationship between business and science is nothing less than an entirely new set of lenses through which to view our organizations."
  —*Library Journal*

"*LEADERSHIP AND THE NEW SCIENCE* IS THE BEST BOOK ON ORGANIZATIONAL LEADERSHIP THAT I HAVE READ IN TEN YEARS. It reminds me of the *LIFE* magazine issue devoted to Picasso. That excited me like few other things. I could not put it down. It captured my imagination and led me to a new plane for living and working. . . . That is what happened with Wheatley's book. It has the same kind of vision and uniqueness. It predicts the future and explains the present like only artists can. This book has helped me impart to others a totally unique kind of trust and courage that works, but seems to go completely contrary to the grain. Simply, it is exquisite."

—Lee M. Hogan, President, Lee Hogan & Associates, and member of the Board of Directors of Associated Consultants International

"WHEATLEY'S INTERPRETATION OF THE 'NEW SCIENCE'—QUANTUM PHYSICS, BIOLOGY, AND CHAOS THEORY—INTO THE ORGANIZING CONCEPTS OF WORK AND PRODUCTIVITY IS PURE GENIUS."

—*The New Leaders*

"THE WORK YOU DO IN THE WORLD IS A WONDER, MEG—so upstream to the way things appear on the surface, but so in harmony with deep-down dynamics of reality. It is a source of great joy to me to think that I have had a small part in the great work you do."

—Parker Palmer, educator and author

"MEG WHEATLEY'S PIONEERING INSIGHTS INTO THE SELF-ORGANIZING NA-TURE OF OUR WORLD HAVE BEEN REMARKABLY WELL SUPPORTED by recent advances in the new sciences. But what really makes *Leadership and the New Science* so enduring is that it offers us a solid place to stand amidst the chaos and complexity. We need this book more than ever."

—Allan Cohen, former Senior Vice President, Zefer, former COO, Waite & Company

"I ADMIRE THE CLARITY, BEAUTY, AND PASSION WITH WHICH YOU TRAVEL YOUR CHOSEN PATH. Without question, you are having considerable impact both here and abroad."

—Robert Tannenbaum, a founder of the field of OD

"I BELIEVE THAT IF THIS BOOK WAS TRANSLATED INTO RUSSIAN, IT WOULD MAKE AN INVALUABLE CONTRIBUTION TO OUR MODERN CULTURE. It has made a huge impression on me not only for its simple, natural words, but the sense of novelty, and singularity of approach. This book for me was certainly a revelation, one of those books that make it worthwhile to study English."

—Mikhail Kutyrev, Russia, former fishing fleet captain

"I'VE SPENT THE LAST FEW DAYS DEVOURING *LEADERSHIP AND THE NEW SCIENCE*, eulogizing about it to my wife, commenting to the friend who lent me the book. 'I'm not sure if it is true, but I really want it to be.' The truth of the matter is that much of what you have written I instinctively know to be true but I have never had quite the words to express it."

    —Steve Clifford, Leadership Development Consultant, England

"I WAS INSPIRED AND PROVOKED BY YOUR WRITINGS to adapt your concepts to the situation of incarcerated individuals. . . . My work is with alcoholics and chemically abusing or addicted individuals. I believe your concepts are a perfect vehicle to reach people who are stuck in denial . . . and your concepts help provide information in a non-threatening and easily understood manner." —Diana Arostegui, Washington

"YOUR BOOK SPEAKS TO ME AND I SPEAK TO IT. I am a practicing lawyer, now almost 75. . . . I always feel the interconnections of everything and the joy of uncertainty and unknowing and am always excited about ideas that I had never thought of and couldn't have by myself." —Dorothy Stulberg, Tennessee

"WHEN WE CAN GET PEOPLE TO PERCEIVE OUR ORGANIZATIONS DIFFERENTLY, THEY ARE BETTER ABLE TO RELATE TO ONE ANOTHER IN THEM. Here is where I find that the 'secular' insights of your book and the 'spiritual' notion of communion are powerful complements to one another. Your simplicity, your directness, and your imagination combine to provide insights that are accessible and compelling."

    —M.B. (Jerry) Handspicker, Professor of Pastoral Theology (Emeritus), Andover
       Newton Theological School

"*LEADERSHIP AND THE NEW SCIENCE* . . . IS ONE OF THOSE BOOKS THAT HELD ME SPELLBOUND with every turn of the page. I felt a keen sense of disappointment when I realized I had come to the last page. You have a gift, Margaret Wheatley, and I am grateful beyond my ability to express that I have been a recipient of it."

    —Maura Jones, North Dakota

". . . I AM VERY ENCOURAGED THAT THERE IS STILL REASON TO BELIEVE ARDENTLY IN THE GREAT RESILIENCE OF THE HUMAN SPIRIT to re-invent itself over and over again, whether we view it fractally or as the first signs of a dawning and long awaited millennium of harmony and happiness for everyone. Perhaps what you are describing is the forerunner not only of new leadership skills and deeper understanding one of another, but also the advent of greater openness of mind and genuine honesty between us all at all levels of communication."

    —Reverend Mary Fourchalk, B.C., Canada

LEADERSHIP
and the
NEW SCIENCE

Other books by Margaret J. Wheatley

A Simpler Way
(with Myron Kellner-Rogers)

Turning to One Another

Finding Our Way

LEADERSHIP
and the
NEW SCIENCE

Discovering Order
in a
Chaotic World

Third Edition

MARGARET J. WHEATLEY

BERRETT-KOEHLER PUBLISHERS, INC.
San Francisco

Berrett-Koehler Publishers, Inc.
235 Montgomery Street, Suite 650
San Francisco, CA 94104-2916
Tel: (415) 288-0260   Fax: (415) 362-2512   www.bkconnection.com

Ordering Information

Quantity sales. Special discounts are available on quantity purchases by corporations, associations, and others. For details, contact the "Special Sales Department" at the Berrett-Koehler address above.

Individual sales. Berrett-Koehler publications are available through most bookstores. They can also be ordered directly from Berrett-Koehler: Tel: (800) 929-2929; Fax: (802) 864-7626; www.bkconnection.com

Orders for college textbook/course adoption use. Please contact Berrett-Koehler: Tel: (800) 929-2929; Fax: (802) 864-7626.

Orders by U.S. trade bookstores and wholesalers. Please contact Ingram Publisher Services, Tel: (800) 509-4887; Fax: (800) 838-1149; E-mail: customer.service@ingrampublisherservices.com; or visit www.ingrampublisherservices.com/Ordering for details about electronic ordering.

Berrett-Koehler and the BK logo are registered trademarks of Berrett-Koehler Publishers, Inc.

Printed in the United States of America

Berrett-Koehler books are printed on long-lasting acid-free paper. When it is available, we choose paper that has been manufactured by environmentally responsible processes. These may include using trees grown in sustainable forests, incorporating recycled paper, minimizing chlorine in bleaching, or recycling the energy produced at the paper mill.

Library of Congress Cataloging-in-Publication Data
Wheatley, Margaret J.
    Leadership and the new science : discovering order in a chaotic world / Margaret J. Weatley. —3rd ed.
        p. cm.
    Includes bibliographical references and index.
    ISBN: 978-1-57675-344-6
    1. Leadership. 2. Organization. 3. Quantum theory. 4. Self-organizing systems. 5. Chaotic behavior in systems. I. Title.
HD57.7.W47 2006
500—dc22   2006042803

11 10 09   10 9 8 7 6 5 4

Designed by Detta Penna
Illustration credits appear on page 194, which constitutes an extension of this copyright page.

Ahó Mitakuye Oyas'in
For all my relations
—*Lakota Nation dedicatory prayer*

My continuing passion is to part a curtain,
that invisible shadow that falls between people,
the veil of indifference to each other's presence,
each other's wonder,
each other's human plight.
—*Eudora Welty*

Contents

Prologue: Maps to the Real World

I have always thought of this book as a collection of intriguing maps, much like those used by the early explorers when they voyaged in search of new lands. Their early maps and commentaries were descriptive but vague, enticing but not fully revealing. They pointed in certain directions, illuminated landmarks, warned of dangers, yet their elusive references and blank spaces served to encourage others to explore and discover. They contained colorful embellishments of places that had struck the discoverer's imagination, yet ignored other important places or contained significant errors. Many early maps contain warnings: "Here there be dragons," or "Regions very imperfectly known." But these maps contained enough knowledge to inspire those who were willing, to dare similar voyages of their own.

The territory that I began mapping when this book was first published in 1992 has now revealed many more of its features. It is the world we live in daily, a world of uncertainty, sudden shifts, and webs of relationships extending around the world. In 1990, as I began to apply the new sciences to the challenges of leadership, I noted that "we live in a time of chaos, as rich in the potential for disaster as for new possibilities." What's ironic is that I *now* look back to 1990 as the good old days, when we had time and space to reflect on ideas, when we had the luxury to think about a new worldview and consider whether we believed it or not. The tone of this book reflects that more spacious era. It is a gentle invitation to become curious, to discover your own questions, to see if your experiences confirm or disconfirm new science, and to engage with me and many others as explorers of this new world only beginning to become visible.

But now my voice of invitation needs to be prefaced by a clear, more insistent voice. Now I am the town crier sounding the alarm. The world has changed. The worldview of the sciences described here is no longer hidden in books. It blares from news reports and blazes across our screens in the terrifying images of these times—wars, terrorism, migrations of displaced people, hurricanes, earthquakes, tsunamis. Chaos and global interconnectedness are part of our daily lives. We try hard to respond to these challenges and threats through our governments, organizations and as individuals, but our actions fail us. No matter what we do, stability and lasting solutions elude us. It's time to realize that we will never cope with this new world using our old maps. It is our fundamental way of interpreting the world—our worldview—that must change. Only such a shift can give us the capacity to understand what's going on, and to respond wisely

I've been out in the world for many years describing the new worldview that science offers us. In my travels, I've met hundreds of thousands of people who have shifted their view and are creating organizations that are adaptive, creative and resilient. Yet many others are more cautious and doubtful. Some people can't be convinced that anything has really changed—the old ways still work fine for them. Others believe that organizations can only function well, especially in times of chaos, by using command and control leadership and hierarchical structures. And many want evidence that these strange new concepts apply 'to the real world.'

Here is the real world as I experience it. It is a world where small groups of enraged people alter the politics of the most powerful nations on earth. It is a world where very slight changes in the temperature of oceans cause violent weather that brings great hardship to people living far from those oceans. It is a world where pandemics kill tens of millions and viruses leap carelessly across national boundaries. It is a world of increased fragmentation where people retreat into positions and identities. It is a world where we have very different interpretations of what's going on, even though we look at the same information. It is a world of constant surprise, where we never know what we'll hear when we turn on the news. It is a world where change is just the way it is.

This dramatic and turbulent world makes a mockery of our plans and predictions. It keeps us on edge, anxious and sleepless. Nothing makes sense anymore. Meaning eludes us. Some offer explanations that this is the end of times or the age of destruction.

Whatever your personal beliefs and experiences, I invite you to consider that we need a new worldview to navigate this chaotic time. We cannot hope to make sense using our old maps. It won't help to dust them off or reprint them in bold colors. The more we rely on them, the more disoriented we become. They cause us to focus on the wrong things and blind us to what's significant. Using them, we will journey only to greater chaos.

Now that I've spent years applying the lens of new science to organizations, communities, governments, nation states, and to myself and family, I can report on the gifts available with a new paradigm. I have discovered insights and explanations about why things are unfolding as they are. I have been inspired to experiment with new ideas and solutions. I feel I am learning how to move more effectively and gracefully through this time.

But I have also discovered how hard it is to surrender a worldview. When scientists confronted this challenge at the beginning of the 20th century, they couldn't accept the world revealed to them in their experiments. They described this new world as strange, puzzling, troubling, bizarre, absurd.

When our worldview doesn't work any longer and we feel ourselves sinking into confusion, of course we feel frightened. Suddenly, there is no ground to stand on. Solutions that worked no longer do. The world appears incomprehensible, chaotic, lacking rationality. We respond to this incoherence by applying old solutions more frantically. We become more rigid about our beliefs. We rely on habit rather than creating new responses. We end up feeling frustrated, exhausted and powerless in the face of so much failure. These frustrations and fears create more aggression. We try to make things work by using brute force rather than intelligence and collaboration.

It was only when scientists were willing to accept their confusion instead of fleeing from it and only when they changed the questions they were asking, only then could they discover the insights and formulations that gave them great new capacity. Once this new worldview came into focus, scientists reengaged with their work with new energy. Wonder, curiosity, and the delight of discovery replaced their fatigue and frustration. I am hopeful that we too can regain our energy and delight by looking at the world of organizations through their worldview. I believe their maps are reliable guides to lands of promise, where human creativity, wisdom and courage can be fully engaged in creating healthy and enduring organizations and societies.

You will find maps of many varieties in this book. Some describe specific new science findings in enough detail that, hopefully, you understand their terrain. Others point out less explored places that need further inquiry. Still others are very detailed, drawing deliberate connections between science and organizational life. And finally, there are records of my personal journey, what I felt and experienced as I brought back questions and insights and applied them in my own work.

Like anyone, my own training and world view bias me. I have focused on the scientific discoveries that intrigued my organizational mind and have ignored many others. This is neither a comprehensive nor a technical guide to new science. It recounts, instead, the voyages I took to but a few of the emerging areas in science, those that enticed me. I was intrigued by three different areas of science: quantum physics, self-organizing systems, and chaos theory. Because I develop the science as I go and relate these three to one another, things will make more sense if you read the chapters in order.

The Introduction and Chapter One introduce all three sciences and the contributions they make to our understanding of the way the world works. These first chapters also provide some initial explanations of sources of order in the universe and speculations on the fears and conditioning that prevent us from appreciating the way that order is created in living systems.

Chapters Two, Three, and Four explore the implications of quantum physics for organizational practices that have, until now, been derived from the seventeenth-century world view of the physics of Isaac Newton. Quantum physics challenges our thinking about observation and perception, participation and relationships, and the influences and connections that work across large and complex systems.

The next chapters, Five and Six, focus on living systems and some new concepts emerging from biology and chemistry. These chapters introduce new ways of understanding disequilibrium and change, and the role disorder plays in creating new possibilities for growth. Information, in our self-organizing universe, is the primary resource necessary to bring things into form. New interpretations are required for there to be new forms or new life. Self-organizing systems demonstrate the ability of all life to organize into systems of relationships that increase capacity. These living systems also demonstrate a different relationship between autonomy and control, showing how a large system maintains itself and grows stronger only as it encourages great amounts of individual freedom.

Chaos theory is the subject of Chapter Seven. Chaos is a necessary process for the creation of new order. This is a world where chaos and order exist as partners, where stasis is never guaranteed nor even desired. I describe several lessons learned form the relationship between these two great forces and how we might think about the workings of chaos in our lives and organizations. I also explore lessons to be learned from fractals—how nature creates its diverse and intricate patterns by the presence of a few basic principles combined with large amounts of individual freedom. And I offer my own observations for how our human need for meaning serves to bring order out of chaos.

Chapter Eight explores life's extraordinary capacity to change, to adapt and grow as required. I explain what I believe to be the underlying processes in living systems that give them this capacity. We have spent several decades attempting to change organizations, communities, nations and each other. We have not been successful in these attempts, or they have resulted in troubling unintended consequences. With so many failures, it seems clear that we need to rethink our basic assumptions about how change happens—for this, life is the best teacher.

In Chapter Nine, I draw together various principles from the sciences to highlight those that can contribute to a "new science" of leadership. This new worldview, with its emerging maps and insights, can teach us how to make sense of this world. Much discovery still awaits us, and I hope many more of you will join in.

And in case you need any more convincing that we need a new worldview to navigate these chaotic times, I have written a new chapter that applies these ideas to "the real world." Chapter Ten uses the lens of new science to bring into focus two of our most critical needs: our ability to respond to disasters and our ability to stop terrorism. For me, the lens of new science illuminates these two challenges brilliantly. It allows us to see things that are invisible with the old lens, the deeper dynamics at play in disaster relief efforts and terrorist networks. Once these dynamics become visible, we have the means to respond far more intelligently to these critical dilemmas. This is the promise of a new paradigm—unsolvable problems suddenly become solvable. We must make use of this promise before the world disintegrates into even more chaos.

The Epilogue closes the book on a more personal and philosophical note. I describe my own discoveries about the nature of this journey and the process of discovery. And I encourage us to understand that we can't make this journey alone—we need good companions, patience, endurance, and courage. After many years and difficult passages, I feel grounded in this new land, nourished by its ideas, and hopeful about its promises. I hope you too will venture forth to make your own discoveries, which you will then offer generously to the rest of us.

To my mind there must be, at the bottom of it all, not an equation, but an utterly simple idea. And to me that idea, when we finally discover it, will be so compelling, so inevitable, that we will say to one another, "Oh, how beautiful. How could it have been otherwise?"

—*John Archibald Wheeler*

Searching for a Simpler Way
to Lead Organizations

I am not alone in wondering why organizations aren't working well. Many of us are troubled by questions that haunt our work. Why do so many organizations feel lifeless? Why do projects take so long, develop ever-greater complexity, yet too often fail to achieve any truly significant results? Why does progress, when it appears, so often come from unexpected places, or as a result of surprises or synchronistic events that our planning had not considered? Why does change itself, that event we're all supposed to be "managing," keep drowning us, relentlessly making us feel less capable and more confused? And why have our expectations for success diminished to the point that often the best we hope for is endurance and patience to survive the frequent disruptive forces in our organizations and lives?

These questions had been growing within me for several years, gnawing away at my work and diminishing my sense of competency. The busier I became with work and the more projects I took on, the greater my questions grew. Until I began a journey.

Like most important journeys, mine began in a mundane place—a Boeing 757, flying soundlessly above America. High in the air as a weekly commuter between Boston and Salt Lake City, with long stretches of reading time broken only by occasional offers of soda and peanuts, I opened my first book on the new science—Fritjof Capra's *The Turning Point*, which describes the new world view emerging from quantum physics. This provided my first glimpse of a new way of perceiving the world, one that comprehended its processes of change, its deeply patterned nature, and its dense webs of connections.

I don't think it accidental that I was introduced to a new way of seeing at 37,000 feet. The altitude only reinforced the message that what was needed was a larger perspective, one that took in more of the whole of things. From that first book, I took off, reading as many new science books as I could find in biology, evolution, chaos theory, and quantum physics. Discoveries and theories of new science called me away from the details of my own field of management and raised me up to a vision of the inherent orderliness of the universe, of creative processes and dynamic, continuous change that still maintained order. This was a world where order and change, autonomy and control were not the great opposites that we had thought them to be. It was a world where change and constant creation were ways of sustaining order and capacity.

I don't believe I could have grasped these ideas if I had stayed on the ground.

During the past several decades, books that relate new science findings for lay readers have proliferated, some more reputable and scientific than others. Of the many I read, some were too challenging, some were too bizarre, but others contained images and information that were breathtaking. I became aware that I was wandering in a realm that created new visions of freedom and possibility, giving me new ways to think about my work. I couldn't always draw immediate connections between science and my dilemmas, but I noticed myself developing a new serenity in response to the questions that surrounded me. I was reading of chaos that contained order; of information as an essential, nourishing element; of systems that fell apart so they could reorganize themselves; and of invisible influences that permeate space and affect change at a distance. These were compelling, evocative ideas, and they gave me hope, even if they did not reveal immediate solutions.

Somewhere—I knew then and believe even more firmly now—there is a simpler way to lead organizations, one that requires less effort and produces less stress than our current practices. For me, this new knowledge is now crystallizing into applications even as I realize that this exploration will take many years. But I no longer believe that organizations are inherently unmanageable in this world of constant flux and unpredictability. Rather, I believe that our present ways of organizing are outmoded, and that the longer we remain entrenched in our old ways, the further we move from those wonderful breakthroughs in understanding that the world of science calls "elegant." The layers of complexity, the sense of things being beyond our control and out of control, are but signals of our failure to understand a deeper reality of organizational life, and of life in general.

We are all searching for this simpler way. In every academic discipline and institution, we live today with questions for which our expertise provides no answers. At the turn of the century, physicists faced the same unnerving confusion. There is a frequently told story about Niels Bohr and Werner Heisenberg, two founders of quantum theory. This version is from *The Turning Point*:

> In the twentieth century, physicists faced, for the first time, a serious challenge to their ability to understand the universe. Every time they asked nature a question in an atomic experiment, nature answered with a paradox, and the more they tried to clarify the situation, the sharper the paradoxes became. In their struggle to grasp this new reality, scientists became painfully aware that their basic concepts, their language, and their whole way of thinking were inadequate to describe atomic phenomena. Their problem was not only intellectual but involved an intense emotional and existential experience, as vividly described by Werner Heisenberg: "I remember discussions with Bohr which went through many hours till very late at night and ended almost in despair; and when at the end of the

discussion I went alone for a walk in the neighboring park I repeated to myself again and again the question: Can nature possibly be so absurd as it seemed to us in these atomic experiments?"

It took these physicists a long time to accept the fact that the paradoxes they encountered are an essential aspect of atomic physics. . . . Once this was perceived, the physicists began to learn to ask the right questions and to avoid contradictions . . . and finally they found the precise and consistent mathematical formulation of [quantum] theory.

. . . Even after the mathematical formulation of quantum theory was completed, its conceptual framework was by no means easy to accept. Its effect on the physicists' view of reality was truly shattering. The new physics necessitated profound changes in concepts of space, time, matter, object, and cause and effect; and because these concepts are so fundamental to our way of experiencing the world, their transformation came as a great shock. To quote Heisenberg again: "The violent reaction to the recent development of modern physics can only be understood when one realizes that here the foundations of physics have started moving; and that this motion has caused the feeling that the ground would be cut from science." (Capra 1983, 76–77)

For the past several years, I have found myself often relating this story to groups of people in organizations everywhere. The story speaks with a chilling familiarity. Each of us recognizes the feelings this tale describes, of being mired in the habit of solutions that once worked yet that are now totally inappropriate, of having rug after rug pulled from beneath us, whether by a corporate merger, reorganization, downsizing, or personal disorientation. But the story also gives great hope as a parable teaching us to embrace our despair as a step on the road to wisdom, encouraging us to sit in the unfamiliar seat of not knowing and open ourselves to radically new ideas. If we bear the confusion, then one day, the story promises, we will begin to see a whole new land, one of bright illumination that will dispel the oppressive shadows of our current ignorance. I still tell Heisenberg's story. It never fails to speak to me from this deep place of reassurance.

I believe that we have only just begun the process of discovering and inventing the new organizational forms that will inhabit the twenty-first century. To be responsible inventors and discoverers, we need the courage to let go of the old world, to relinquish most of what we have cherished, to abandon our interpretations about what does and doesn't work. We must learn to see the world anew. As Einstein is often quoted as saying: No problem can be solved from the same consciousness that created it.

There are many places to search for new answers in a time of paradigm shifts. For me, it was appropriate that my inquiry led back to the natural sciences, re-connecting me to an earlier vision of myself. At fourteen, I aspired to be a space biologist and carried heavy astronomy texts on the New York subway to weekly classes at the Hayden Planetarium. These texts were far too dense for me to understand, but I carried them anyway because they looked so impressive. My abilities in biology were better founded, and I began college majoring in biology, but my encounters with advanced chemistry ended that career, and I turned to the greater ambiguity of the social sciences. Like many social scientists, I am at heart a lapsed scientist, still hoping the world will yield up its secrets to me.

But my focus on science is more than a personal interest. Each of us lives and works in organizations designed from Newtonian images of the universe. We manage by separating things into parts, we believe that influence occurs as a direct result of force exerted from one person to another, we engage in complex planning for a world that we keep expecting to be predictable, and we search continually for better methods of objectively measuring and perceiving the world. These assumptions, as I explain in Chapter Two, come to us from seventeenth-century physics, from Newtonian mechanics. They are the basis from which we design and manage organizations, and from which we do research in all of the social sciences. Intentionally or not, we work from a world view that is strongly anchored in the natural sciences.

But the science has changed. If we are to continue to draw from science to create and manage organizations, to design research, and to formulate ideas about organizational design, planning, economics, human motivation, and change processes (the list can be much longer), then we need to at least ground our work in the science of our times. We need to stop seeking after the universe of the seventeenth century and begin to explore what has become known to us during the twentieth century. We need to expand our search for the principles of organization to include what is presently known about how the universe organizes.

The search for the lessons of new science is still in progress, really in its infancy; but what I hope to convey in these pages is the pleasure of sensing those first glimmers of a new way of thinking about the world and its organizations. The light may be dim, but its potency grows as the door cracks wider and wider. Here there are scientists who write about natural phenomena with a poetry and a clarity that speak to dilemmas we find in organizations. Here there are new images and metaphors for thinking about our own organizational experiences. This is a world of wonder and not knowing, where many scientists are as awestruck by what they see as were the early explorers who marveled at new continents. In this realm, there is a new kind of freedom, where it is more rewarding to explore than to reach conclusions, more satisfying to wonder than to know, and more exciting to search than to stay put. Curiosity, not certainty, becomes the saving grace.

This is not a book filled with conclusions, cases, or exemplary practices. It is deliberately *not* that kind of book, for two reasons. First, I don't believe that organizations are ever changed by imposing a model developed elsewhere. So little transfers to, or inspires, those trying to work at change in their own organizations. In every organization, we need to look internally, to see one another as the critical resources on this voyage of discovery. We need to learn how to engage the creativity that exists everywhere in our organizations. Second, the new physics cogently explains that there is no objective reality out there waiting to reveal its secrets. There are no recipes or formulas, no checklists or expert advice that describe "reality." If context is as crucial as the science explains, then nothing really transfers; everything is always new and different and unique to each of us. We must engage with each other, experiment to find what works for us, and support one another as the true inventors that we are.

This book attempts to be true to that new vision of reality, where ideas and information are but half of what is required to evoke reality. The creative possibilities of the ideas represented here depend on your engagement with them. I assigned myself the task of presenting material to provoke and engage you, knowing that your experience with these pages will produce different ideas, different hopes, and different experiments than mine. It is not important that we agree on one expert interpretation or one best practice. That is not the nature of the universe in which we live. We inhabit a world that co-evolves as we interact with it. This world is impossible to pin down, constantly changing, and infinitely more interesting than anything we ever imagined.

Though the outcomes to be gained from reading this book are unique to each of you, the ideas I have chosen to think about focus on the meta-issues that concern those of us who work in organizations: Where is order to be found? How do complex systems change? How do we create structures that are flexible and adaptive, that enable rather than constrain? How do we simplify things without losing what we value about complexity? How do we resolve personal needs for autonomy and growth with organizational needs for prediction and accountability?

The new science research referred to in this book comes from the disciplines of physics, biology, and chemistry, and from theories of evolution and chaos that span several disciplines. Each chapter inquires into metaphorical links between certain scientific perspectives and organizational phenomena, but it may be useful first to say something about the direction of new science.

Scientists in many different disciplines are questioning whether we can adequately explain how the world works by using the machine imagery emphasized in the seventeenth century by such great geniuses as Sir Isaac Newton and René Descartes. This machine imagery leads to the belief that studying the parts is the key to understanding the whole. Things are taken apart, dissected literally or figuratively (as we have done with business functions, academic disciplines, areas of specialization, human body parts), and then put back together without any significant loss. The assumption is that the more we know about the workings of each piece, the more we will learn about the whole.

Newtonian science is also materialistic—it seeks to comprehend the world by focusing on what can be known through our physical senses. Anything *real* has visible and tangible physical form. In the history of physics and even to this day, many scientists keep searching for the basic "building blocks" of matter, the physical forms from which everything originates.

One of the first differences between new science and Newtonianism is a focus on holism rather than parts. Systems are understood as whole systems, and attention is given to *relationships within those networks*. Donella Meadows, an ecologist and author, quotes an ancient Sufi teaching that captures this shift in focus: "You think because you understand *one* you must understand *two*, because one and one makes two. But you must also understand *and*" (1982, 23). When we view systems from this perspective, we enter an entirely new landscape of connections, of phenomena that cannot be reduced to simple cause and effect, or explained by studying the parts as isolated contributors. We move into a land where it becomes critical to sense the constant workings of dynamic processes, and then to notice how these processes materialize as visible behaviors and forms.

Explorations into the subatomic world began early in this century, creating the dissonance described in Heisenberg's story. In physics, therefore, the search for radically new models now has a long and somewhat strange tradition. The strangeness lies in the pattern of discovery that characterized many of the major discoveries in quantum mechanics: "A lucky guess based on shaky arguments and absurd ad hoc assumptions gives a formula that turns out to be right, though at first no one can see why on earth it should be" (March 1978, 3). I delight in that statement of scientific process. It gives me hope that we might all approach discovery differently, hope that we can move away from the plodding, deadening character of so many research and planning activities.

The quantum mechanical view of reality startles us out of common notions of what is real. Even to scientists, it is admittedly bizarre. In the quantum world, *relationship* is the key determiner of everything. Subatomic particles come into form and are observed only as they are in relationship to something else. They do not exist as independent "things." There are no basic "building blocks." Quantum physics paints a strange yet enticing view of a world that, as Heisenberg characterized it, "appears as a complicated tissue of events, in which connections of different kinds alternate or overlap or combine and thereby determine the texture of the whole" (1958, 107). These unseen *connections* between what were previously thought to be separate entities are the fundamental ingredient of all creation.

In other disciplines, especially biology, nonmechanistic models are only beginning to be replaced by more holistic, dynamic ones. Traditional mechanistic thinking still prevails in the field of molecular biology and most work in genetics. But many scientists now seek to understand *life as life,* moving away from machine imagery. For example, in *The Web of Life* (1996), Fritjof Capra presents a new synthesis of the science of living systems, drawing together scientific discoveries and theories from many branches of science. Capra's synthesis reveals processes that are startlingly different from the mechanistic ones that had been used to explain life.

Similar shifts in understanding have appeared in the field of human health. In holistic treatments, the body is viewed as an integrated system rather than as a collection of discrete parts. Some biologists offer the perspective that what we thought of as discrete systems (such as the immune, endocrine, and neurological systems) are better understood as one system, totally inter-dependent in their functioning (see Pert and Chopra 1997).

And at the grandest level of scale, looking at the earth as a whole, is the Gaia theory, first proposed by James Lovelock. There is increasing support for his hypothesis that the earth is a self-regulating system, a planetary community of interdependent systems that together create the conditions which make life possible (see Lovelock 1988, Margulis 1998).

In biology, so many fundamental reformulations of prevailing theories are occurring—in evolution, animal behavior, ecology, physiology—that Ernst Mahr, a noted chronicler of biological thought, stated that a new philosophy of biology is needed (1988). What is being sought, comments biologist Steven Rose, is a biology that is more holistic and integrative, a "science that is adult enough to rejoice in complexity" (1997, 133).

In chemistry, Ilya Prigogine won the Noble Prize in 1977 for work that demonstrates how certain chemical systems reorganize themselves into greater *order* when confronted with changes in their environment. In the older, mechanistic models of systems, change and disturbances signaled trouble. These disruptions would only speed up the inevitable decline that was the fate of all systems. But Prigogine's work offered a new and more promising future. He demonstrated that any open system has the capacity to respond to change and disorder by reorganizing itself at a higher level of organization. Disorder becomes a critical player, an ally that can provoke a system to self-organize into new forms of being. As we leave behind the machine model of life and look more deeply into the dynamics of living systems, we begin to glimpse an entirely new way of understanding fluctuations, disorder, and change.

New understandings of change and disorder have also emerged from chaos theory. Work in this field has led to a new appreciation of the relationship between order and chaos. These two forces are now understood as mirror images, two states that contain the other. A system can descend into chaos and unpredictability, yet within that state of chaos the system is held within boundaries that are well-ordered and predictable. Without the partnering of these two great forces, no change or progress is possible. Chaos is necessary to new creative ordering. This revelation has been known throughout time to most human cultures; we just needed the science to help us remember it.

New science is also making us more aware that our yearning for freedom and simplicity is one we share with all life. In many examples, scientists now describe how order and form are created not by complex controls, but by the presence of a few guiding formulas or principles repeating back on themselves through the exercise of individual freedom. The survival and growth of systems that range in size from large ecosystems down to the smallest microbial colonies are sustained by a few key principles that express the system's overall identity combined with high levels of autonomy for individuals within that system.

The world described by new science is changing our beliefs and perceptions in many areas, not just those of science. New science ideas have crept into almost every discipline, including my own field of organizational theory. I can see the influence of science if I look at those problems that plague us most in organizations and how we are reformulating them. Leadership, an amorphous phenomenon that has intrigued us since people began organizing, is being examined now for its relational aspects. Few if any theorists ignore the complexity of relationships that contribute to a leader's effectiveness. Instead, there are more and more studies on partnership, followership, empowerment, teams, networks, and the role of context.

Relational issues appear everywhere I look. Ethical and moral questions are no longer fuzzy religious concepts but key elements in the relationship any organization has with colleagues, stakeholders, and communities. At the personal level, many authors write now on our interior relationship with our spirit, soul, and life's purpose. Ecological writers stress the relationship that exists not only between us and all beings in our environment, but between us and future generations. If the physics of our time is revealing the primacy of relationships, is it any wonder that we are beginning to rethink our major issues in more relational terms?

In motivation theory, attention is shifting from the use of external rewards to an appreciation for the intrinsic motivators that give us great energy. We are refocusing on the deep longings we have for community, meaning, dignity, purpose, and love in our organizational lives. We are beginning to look at the strong emotions of being human, rather than segmenting ourselves by believing that love doesn't belong at work, or that feelings are irrelevant in the organization. There are many attempts to leave behind the view that predominated in the twentieth century, when we believed that organizations could succeed by confining workers to narrow roles and asking only for very partial contributions. As we let go of the machine model of organizations, and workers as replaceable cogs in the machinery of production, we begin to see ourselves in much richer dimensions, to appreciate our wholeness, and, hopefully, to design organizations that honor and make use of the great gift of who we humans are.

The impact of vision, values, and culture occupies a great deal of organizational attention. We see their effects on organizational vitality, even if we can't define why they are such potent forces. We now sense that some of the best ways to create continuity and congruence in the midst of turbulent times are through the use not of controls, but of forces that are invisible yet palpable. Many scientists now work with the concept of fields—invisible forces that occupy space and influence behavior. I have played with the notion that organizational vision and values act like fields, unseen but real forces that influence people's behavior. This is quite different from more traditional notions that vision is an evocative message about some desired future state delivered by a charismatic leader.

Our concept of organizations is moving away from the mechanistic creations that flourished in the age of bureaucracy. We now speak in earnest of more fluid, organic structures, of boundaryless and seamless organizations. We are beginning to recognize organizations as whole systems, construing them as "learning organizations" or as "organic" and noticing that people exhibit self-organizing capacity. These are our first journeys that signal a growing appreciation for the changes required in today's organizations. My own experience suggests that we can forego the despair created by such common organizational events as change, chaos, information overload, and entrenched behaviors if we recognize that organizations are living systems, possessing the same capacity to adapt and grow that is common to all life.

Some believe that there is a danger in playing with science and abstracting its metaphors because, after a certain amount of stretch, the metaphors lose their relationship to the tight scientific theories that gave rise to them. But others would argue that all science is metaphor, a hypothetical description of how to think of a reality we can never fully know. In seeking to play with the rich images coming out of new science, I share the sentiments of physicist Frank Oppenheimer: "If one has a new way of thinking, why not apply it wherever one's thought leads to? It is certainly entertaining to let oneself do so, but it is also often very illuminating and capable of leading to new and deep insights" (Cole 1985, 2).

One learns to hope that nature possesses an order that one may aspire to comprehend.

—C. N. Yang

Chapter 1

Discovering an Orderly World

It has taken us a long while to get here—a nine-mile hike up a gradual ascent over rocky paths. My horse, newly trained to pack equipment and still an amateur, has bumped against my back, bruised my heels, and finally, unavoidably, stepped on my toe, smashing it against the inside of my boot. But it's been worth it. Here are the American Rockies at their clichéd best. The stream where I sit soaking my feet glistens on for miles I can't see, into green grasses that bend to the wind. There are pine trees, mountains, hawks, and off at the far edge of the meadow a moose who sees us and moves to hide her great girth behind a tree that is only four inches wide. The tree extends just to the edge of each eyeball. We laugh, but I suspect there's a lesson in it for all of us.

For months, I have been studying process structures—things that sustain their identity over time yet are not locked rigidly into any one physical form. This stream that swirls around my feet is the most beautiful one I've encountered. Because it is vacation, I resist thinking too deeply about this stream, but as I relax into its flow, images stir and gently whorl the surface.

Finally, I ask directly: What is it that streams can teach me about organizations? I am attracted to the diversity I see, to these swirling combinations of mud, silt, grass, water, rocks. This stream has an impressive ability to adapt, to change the configurations, to let the power shift, to create new structures. But behind this adaptability, making it all happen, I think, is the water's need to flow. Water answers to gravity, to downhill, to the call of ocean. The forms change, but the mission remains clear. Structures emerge, but only as temporary solutions that facilitate rather than interfere. There is none of the rigid reliance that I have learned in organizations on single forms, on true answers, on past practices. Streams have more than one response to rocks; otherwise, there'd be no Grand Canyon. Or Grand Canyons everywhere. The Colorado river realized there were many ways to find ocean other than by staying broad and expansive.

Organizations lack this kind of faith, faith that they can accomplish their purposes in varied ways and that they do best when they focus on intent and vision, letting forms emerge and disappear. We seem hypnotized by structures, and we build them strong and complex because they must, we believe, hold back the dark forces that threaten to destroy us. It's a hostile world out there, and organizations, or we who create them, survive only because we build crafty and smart—smart enough to defend ourselves from the natural forces of destruction. Streams have a different relationship with natural forces. With sparkling confidence, they know that their intense yearning for ocean will be fulfilled, that nature creates not only the call, but the answer.

Many of the organizations I experience are impressive fortresses. The language of defense permeates them: in CYA memo-madness; in closely guarded secrets and locked personnel files; in activities defined as "campaigns," "skirmishes," "wars," "turf battles," and the ubiquitous phrases of sports that describe everything in terms of offense and defense. Many organizations feel they have to defend themselves even against their employees with regulations, guidelines, time clocks, and policies and procedures for every eventuality. One organization I worked in welcomed its new employees with a list of twenty-seven offenses for which they could be summarily fired—and the assurance that they could be fired for other reasons as well. Some organizations have rigid chains of command to keep people from talking to anyone outside their department, and in most companies, protocols define who can be consulted, advised, or criticized. We are afraid of what would happen if we let these elements of the organization recombine, reconfigure, or speak truthfully to one another. We are afraid that things will fall apart.

This need to hold the world together, these experiences of fright and fragility, are so pervasive that I wondered about the phenomenon long before I came upon this teacher stream. Fear that is everywhere must come to us from somewhere. But where? In modern Western thought, I believe one source is our fuzzy understanding of concepts that gained strength from seventeenth-century science. Three centuries ago, when the world was imagined as an exquisite machine set in motion by God—a closed system with a watchmaker father who then left the shop—the concept of entropy entered our collective consciousness. Machines wear down; they eventually stop. In the poet Yeats' phrase, "Things fall apart; the center cannot hold, mere anarchy is loosed upon the world." This is a universe, we feel, that cannot be trusted with its own processes for growth and rejuvenation. If we want progress, then we must provide the energy to reverse decay. By sheer force of will, because we are the planet's intelligence, we will make the world work. We will resist death.

What a fearful posture this has been! Something Atlas only imagined, it has gone on so long. It is time to stop now. It is time to take the world off our shoulders, to lay it gently down and look to it for an easier way. It is not only streams that have something to teach us. Lessons are everywhere. But the question is key. If not with us, then where are the sources of order to be found?

I believe nature offers abundant displays of order and clear lessons for how to achieve it. Despite the experience of fluctuations and changes that disrupt our plans, the world is inherently orderly. It continues to create systems of great scope, capacity, and diversity. And fluctuation and change are essential to the process by which order is created.

Life is about creation. This ability of life to create itself is captured in a strange-sounding new word, *autopoiesis* (from Greek, meaning self-production or self-making). Autopoiesis is life's fundamental process for creating and renewing itself, for growth and change. A living system is a network of processes in which every process contributes to all other processes. The entire network is engaged together in producing itself (Capra 1996, 99). This process is not limited to one type of organism—it describes life itself. As described by systems scientist Erich Jantsch, any living system is "a never resting structure that constantly seeks its own self-renewal" (1980, 10). And this description defines a paradox that is important to note when we think about change: A living system produces itself; it will change in order to preserve that self. Change is prompted only when an organism decides that changing is the only way to maintain itself.

There is another important paradox in living systems: Each organism maintains a clear sense of its individual identity *within* a larger network of relationships that helps shape its identity. Each being is noticeable as a separate entity, yet it is simultaneously part of a whole system. While we humans observe and count separate selves, and pay a great deal of attention to the differences that seem to divide us, in fact we survive only as we learn how to participate in a web of relationships. Autopoiesis describes a very different universe, one in which all organisms are capable of creating a "self" through their intimate engagement with all others in their system. This is not a fragile, fragmented world that needs us to hold it together. This is a world rich in processes that support growth and coherence through paradoxes that we need to contemplate.

In chemistry, Ilya Prigogine's prize-winning work also teaches a paradoxical truth, that disorder can be the source of new order. Prigogine coined the term "dissipative structures" for these newly discovered systems to describe their contradictory nature. Dissipation describes loss, a process of energy gradually ebbing away, while structure describes embodied order. Prigogine discovered that the dissipative activity of loss was necessary to create new order. Dissipation didn't lead to the death of a system. It was part of the process by which the system let go of its present form so that it could reorganize in a form better suited to the demands of its changed environment.

Prigogine's work has helped explain a long-standing contradiction of Western science. If, as science believed, entropy is the rule, then why does life flourish? Why does life result in newness and evolution, not deterioration and disintegration?

In a dissipative structure, anything that disturbs the system plays a crucial role in helping it self-organize into a new form of order. Whenever the environment offers new and different information, the system chooses whether to accept that provocation and respond. This new information might be only a small difference from the norm. But if the system pays attention to this information, it brings the information inside, and once inside that network, the information grows and changes. If the information becomes such a large disturbance that the system can no longer ignore it, then real change is at hand. At this moment, jarred by so much internal disturbance and far from equilibrium, the system will fall apart. In its current form, it cannot deal with the disturbance, so it dissolves. But this disintegration does not signal the death of the system. If a living system can maintain its identity, it can self-organize to a higher level of complexity, a new form of itself that can deal better with the present.

In this way, dissipative structures demonstrate that *disorder* can be a source of new *order*, and that growth appears from disequilibrium, not balance. The things we fear most in organizations—disruptions, confusion, chaos—need not be interpreted as signs that we are about to be destroyed. Instead, these conditions are necessary to awaken creativity. Scientists in this newly understood world describe the relation of disorder to order as "order out of chaos" or "order through fluctuation" (Prigogine and Stengers, 1984). These are new principles that highlight the dynamics between chaos and creativity, between disruption and growth.

At the quantum level of reality, the paradoxes grow even larger. At the subatomic level, change happens in jumps, beyond any power of precise prediction. Quantum physicists speak in terms of probabilities, not prediction. They can calculate the probable moment and location of a quantum leap, but not exactly. Newtonian physics operates with a different belief—that the world *does* behave in deterministic ways. (This assumption has been challenged by Prigogine's recent work; see 1998.)

The quantum world also challenges beliefs about objective measurement, for at the subatomic level the observer cannot observe anything without interfering or, more precisely, participating in its creation. The strange qualities of the quantum world have shaken prevailing scientific beliefs in determinism, predictability, and control. At first glance then, quantum physics doesn't seem to volunteer concepts that aid us in our search for a more orderly universe. But the impossibility of exact predictions at the quantum level is not a result of inherent disorder. Instead, the behaviors observed are a result of quantum interconnectedness, of a deep and intimate order. There is a constant weaving of relationships, of energies that merge and change, of momentary ripples that become noticeable within a seamless fabric. There is so much order that our attempts to separate out discrete events create the appearance of disorder.

Order has been found even in the event that historically has meant absolute disorder—chaos. Chaos theory has given us images of "strange attractors"— computer-generated pictures of swirling motion that trace the evolution of a system. A system is defined as chaotic when it becomes impossible to know what it will do next. The system never behaves the same way twice. But as chaos theory shows, if we look at such a system over time, it demonstrates an inherent orderliness. Its wild gyrations are held within an invisible boundary.

The system holds order within it, and reveals this self-portrait as a beautiful pattern, its strange attractor (see the color section and page 117).

Throughout the universe, then, order exists within disorder and disorder within order. We have always thought that disorder was the absence of the natural state of order, seen in the word itself: dis-order. But do we believe this? Is chaos an irregularity, or is order just a lucky moment grabbed from natural disorder? We've been taught to see things as separate states: One needs to be normal, the other exceptional. Yet as we move into this new territory where paradox is a distinguishing feature, we can see that what is happening is a dance—of chaos and order, of change and stability. Just as in the timeless image of yin and yang, we are dealing with complementarities that only look like polarities. Neither one is primary; both are absolutely necessary. When we observe growth, we observe the results of the dance.

One systems scientist said that a system is *a set of processes* that are made visible in temporary structures. These living structures are in no way similar to the solid structures we build. The structures of life are transient; they are capable of changing if needed: "Caterpillar and butterfly, for example, are two temporarily stabilized structures in the coherent evolution of one and the same system" (Jantsch 1980, 6). The system continues to develop, to release itself from the old and find new structures as they are required.

While we have lusted for order in organizations, we have failed to understand where to find it. We have seen order reflected in the structures we build, whether they be bright mirror-glass buildings, dazzling charts, or plans begun on paper napkins. These structures take so much time, creativity, and attention that it is hard not to want them to be permanent. It is hard to welcome disorder as a full partner in the search for order when we have expended so much effort to bar it from the gates. I find myself challenged by this new land of evolving form, of structures that come and go, of bearings gained not from the rigid artifacts of organization charts and job descriptions, but from directions arising out of deep, natural processes of growth and self-renewal. This is not an easy land to inhabit, not an easy world in which to place faith, except that we're already living with the evidence that supports it—this wonderfully diverse and creative planet. And all of us, even in rigid organizations, have experienced self-organization, times when we recreate ourselves, not according to some idealized plan, but because the environment demands it. We let go of our old form and figure out how best to organize ourselves in new ways.

When I think about the work experiences I cherish most, I see such self-organization. In the interest of getting things done, our roles and tasks moved with such speed that they blurred to nothing. We were too engaged with the work to worry about defining accountabilities or roles. We all felt accountable for figuring out what worked and implementing it quickly. When people speak of informal leadership, they describe a similar experience—how people create the leadership that best responds to their needs at the time. We may fail to honor these leaders more formally, trapped as we are in our beliefs about hierarchy and power, but we always know who the real leader is and why we are willing to follow. Max De Pree, former CEO of Herman Miller, calls this "roving leadership, the indispensable people in our lives who are there when we need them" (1989, 41–42). They emerge from the group, not by self-assertion, but because they make sense, given what the group and individuals need so that they can survive and grow. Organization consultant Jill Janov states that leadership is best thought of as a behavior, not a role. We always need leaders, but this need can be satisfied by many different people, depending on the context (Janov 1994).

All this time, we have created trouble for ourselves in organizations by confusing control with order. This is no surprise, given that for most of its written history, leadership has been defined in terms of its control functions. Lenin spoke for many leaders when he said: "Freedom is good, but control is better." And our quest for control has been oftentimes as destructive as was his.

If people are machines, seeking to control us makes sense. But if we live with the same forces intrinsic to all other life, then seeking to impose control through rigid structures is suicide. If we believe that there is no order to human activity except that imposed by the leader, that there is no self-regulation except that dictated by policies, if we believe that responsible leaders must have their hands into everything, controlling every decision, person, and moment, then we cannot hope for anything except what we already have—a treadmill of frantic efforts that end up destroying our individual and collective vitality.

What if we could reframe the search? What if we stopped looking for control and began, in earnest, the search for order? Order we will find in places we never thought to look before—all around us in nature's living, dynamic systems. In fact, once we begin to look into nature with new eyes, the teachers are everywhere.

I looked again at the moose, staring intently into that narrow beam of tree. Our search for safety, our belief that we can control our organizations by the structures we impose, is no less foolish. As long as we stare cross-eyed at that tree, we can't see all around us the innate processes of living systems that are there to help create the order we crave.

Yet it is hard to step away from that tree. It is hard to open ourselves to a world of inherent orderliness. "In life, the issue is not control, but dynamic connectedness," Jantsch writes (1980, 196). I want to act from that knowledge. I want to trust in this universe so much that I give up playing God. I want to stop struggling to hold things together. I want to experience such security that the concept of "allowing"—trusting that the appropriate forms will emerge—ceases to be scary. I want to surrender my fear of the universe and join with everyone I know in an organization that opens willingly to its environment, participating gracefully in the unfolding dance of order.

For fragmentation is now very widespread, not only throughout society, but also in each individual; and this is leading to a kind of general confusion of the mind, which creates an endless series of problems and interferes with our clarity of perception so seriously as to prevent us from being able to solve most of them . . . . The notion that all these fragments are separately existent is evidently an illusion, and this illusion cannot do other than lead to endless conflict and confusion.

—David Bohm

Chapter 2

Newtonian Organizations in a Quantum Age

I sit in a room without windows, participating in a ritual etched into twentieth-century tribal memory. I have been here thousands of times before, literally. I am in a meeting, trying to solve a problem. Using whatever analytic tool somebody has just read about or been taught at their most recent training experience, we are trying to come to grips with a difficult situation. Perhaps it is poor employee morale or productivity. Or production schedules. Or the redesign of a function. The topic doesn't matter. What matters is how familiar and terrible our process is for coming to terms with the complaint.

The room is adrift in flip-chart paper—clouds of lists, issues, schedules, plans, accountabilities crudely taped to the wall. They crack and rustle, fall loose, and, finally, are pulled off the walls, tightly rolled, and transported to some innocent secretary, who will litter the floor around her desk and peering down from her keyboard, will transcribe them and e-mail them to us. They will appear on our desktops hours or days later, faint specters of commitments and plans, devoid of even the little energy and clarity that sent the original clouds— poof—up onto the wall. They will drift into our calendars and onto individual "to do" lists, lists already fogged with confusion and inertia. Whether they get done or not, they will not solve the problem.

I am weary of the lists we make, the time projections we spin out, the breaking apart and putting back together of problems. It does not work. The lists and charts we make do not capture experience. They only tell of our desire to control a reality that is slippery and evasive and perplexing beyond comprehension. Like bewildered shamans, we perform rituals passed down to us, hoping they will perform miracles. No new wisdom teacher has appeared to show us how to live more wisely in this universe. Our world grows more disturbing and mysterious, our failures to predict and control leer back at us from many places, yet to what else can we turn? If the world is not a machine, then our approaches cannot work. But then, where are we?

The search for new shamans has begun in earnest. Our seventeenth-century organizations are crumbling. We have prided ourselves, in all these centuries since Newton and Descartes, on the triumphs of reason, on the absence of magic. Yet we, like the best magicians of old, have been hooked on manipulation. For three centuries, we've been planning, predicting, and analyzing the world. We've held on to an intense belief in cause and effect. We've raised planning to the highest of priestcrafts and imbued numbers with absolute power. We look to numbers to describe our economic health, our productivity, our physical well-being. We've developed graphs and charts and plans to take us into the future, revering them as ancient mariners did their chart books. Without them, we'd be lost, adrift among the dragons. We have been, after all, no more than sorcerers, the master magicians of our time.

The universe that Sir Isaac Newton described was a seductive place. As the great clock ticked, we grew smart and designed the age of machines. As the pendulum swung with perfect periodicity, it prodded us on to new discoveries. As the Earth circled the sun (just like clockwork), we grew assured of the role of determinism and prediction. We absorbed expectations of regularity into our very beings. And we organized work and knowledge based on our beliefs about this predictable universe.

It is interesting to note just how Newtonian most organizations are. The machine imagery of the cosmos was translated into organizations as an emphasis on material structure and multiple parts. Responsibilities have been organized into functions. People have been organized into roles. Page after page of organizational charts depict the workings of the machine: the number of pieces, what fits where, who the most important pieces are. The 1990s revealed these deeply embedded beliefs about organizations as machines when "reengineering" became the dominant solution for organizational ills. Its costly failures were later acknowledged to have stemmed in large part from processes and beliefs that paid no attention to the human (or living) dimensions of organizational life (see Hammer 1995). William Bygrave, a physicist turned organizational theorist, comments on how many management theorists either were engineers or admired that profession, from Chandler to Porter—a lineage that continues to the present. There has been a close connection, he writes, between their engineering training and their attempts to create a rational, structured approach to organizations (1989, 16).

This reduction into parts and the proliferation of separations has characterized not just organizations, but everything in the Western world during the past three hundred years. We broke knowledge into separate disciplines and subjects, built offices and schools with divided spaces, developed analytic techniques that focus on discrete factors, and even counseled ourselves to act in fragments, to use different "parts" of ourselves in different settings.

In organizations, we focused attention on structure and organizational design, on gathering extensive numerical data, and on making decisions using sophisticated mathematical formulas. We've spent years moving pieces around, building elaborate models, contemplating more variables, creating more precise forms of analysis. Until recently we really believed that we could study the parts, no matter how many of them there were, to arrive at knowledge of the whole. We have reduced and described and separated things into cause and effect, and drawn the world in lines and boxes.

A world based on machine images is a world described by boundaries. In a machine, every piece knows its place. Likewise, in Newtonian organizations, we've drawn boundaries everywhere. We've created roles and accountabilities, specifying lines of authority and limits to responsibilities. We have drawn boundaries around the flow of experience, fragmenting whole networks of interaction into discrete steps. We study variables as separate and well-bounded, even when we attempt to account for some of their interactions through complex statistical techniques. Information is arrayed in two-dimensional charts and graphs that chunk up the world. Charts tell us about market share, employee opinions, customer ratings. We have even come to think of power—an elusive, energetic force if ever there was one—as a bounded resource, defined as "my share of the pie."

These omnipresent boundaries create a strong sense of solidity; we use them to both protect and define us. Boundaries make it possible to know the difference between one thing and another. "The whole corpus of classical physics," writes Danah Zohar in *The Quantum Self*, "and the technology that rests on it is about the separateness of things, about constituent parts and how they influence each other across their separateness" (1990, 69). Classical physics studies a world of things and how influences work across the separations. In a world of things, there are well-defined edges; it is possible to tell where one stops and the other begins, to stand outside something and observe it without interfering. The "thing" view of the world, therefore, has led to a belief in scientific objectivity. And we prospered with this belief for many centuries, working well in a world of you–me, inside–outside, here–thereness.

A vast and complex machine had been entrusted to our care. We searched to know the mind of the clock maker, even as he receded deep into the distance. We made some assumptions about him (gender was never a question). He was infinitely rational, his works were totally predictable, and a few simple laws would reveal what made everything work. Reductionist thinking seduced us into believing that, eventually, we would figure everything out. We would control it all, even life and death. Science displaced God. "Chaos was merely complexity so great," Briggs and Peat comment, "that in practice scientists couldn't track it, but they were sure that in principle they might one day be able to do so. When that day came there would be no chaos, so to speak, only Newton's laws. It was a spellbinding idea" (1989, 22).

In physics, this search for the ultimate laws has led to work on a unified theory, now dubbed the "theory of everything" (see Davies and Brown 1988). Some scientists still believe they will discover the essential secrets of life and be able to control all aspects of existence. While some in management dream of similar levels of control, their desire for prediction has led to less noteworthy results. True simplicity has been confused with a propensity for simplistic exhortations and mindless aphorisms about what makes for a well-run organization.

It has not been easy living in this machine universe. A mechanical world feels distinctly anti-human. As Zohar eloquently describes it, "Classical physics transmuted the living cosmos of Greek and medieval times, a cosmos filled with purpose and intelligence and driven by the love of God for the benefit of humans, into a dead, clockwork machine. . . . Things moved because they were fixed and determined; cold silence pervaded the once-teeming heavens. Human beings and their struggles, the whole of consciousness, and life itself were irrelevant to the workings of the vast universal machine" (1990, 18).

The removal of human experience from the scientific world view had one other surprising consequence. Though scientists had engaged in a successful dialogue with nature, as Prigogine and Stengers describe it, an unexpected outcome of their work "was the discovery of a silent world. This is the paradox of classical science. It revealed a dead, passive nature, a nature that behaves as an automaton which, once programmed, continues to follow the rules inscribed in the program. In this sense, the dialogue with nature isolated humans from nature instead of bringing them closer to it . . . . It seemed that science debased everything it touched" (1984, 6).

Loneliness pervaded not only science, but all Western culture. In America, we raised individualism to its highest expression, each of us protecting our boundaries, asserting our rights, creating a world that, Bellah et al. writes, "leaves the individual suspended in glorious, but terrifying, isolation" (1985, 6).

In science, the beginning of the twentieth century heralded the end of the hegemony of Newtonian thinking. Discoveries of a strange world at the subatomic level could not be explained by Newton's laws, and the path was open for new ways of comprehending the universe. Newtonian mechanics still contribute greatly to scientific advances, but a new and different science is required now to explain many phenomena. Quantum mechanics does not describe a clock-like universe. It tells a very different story:

> Most of the other giant steps in our understanding of nature were really *evolutionary* in that they sprang from previously established foundations: facts were reorganized, or connected in new ways, or seen in a different context. Quantum theory, however, broke away completely from those foundations; it dove right off the end. It could not (cannot) adequately be described in metaphors borrowed from our previous view of reality because many of those metaphors no longer apply. But the net result has not been to obscure reality or make the nature of things more elusive and murky. On the contrary, most physicists would agree that what quantum theory has brought to science is exactly the opposite—concreteness and clarity. (Cole 1985, 106)

Though it may be concrete and clear, the quantum world is weird, even to scientists. Two of its most outstanding theoreticians made strong comments about this. Niels Bohr warns that "Anyone who is not shocked by quantum theory has not understood it." And Erwin Shroedinger, reacting to some of its puzzles, says, "I don't like it and I'm sorry I ever had anything to do with it" (in Gribbin 1984, 5; frontispiece).

But the quantum world is not just weird and fascinating. As more of us contemplate these strange behaviors at the subatomic level, I believe we are given potent images that can enrich our lives at the macro level. Quantum imagery challenges so many of our basic assumptions, including our understanding of relationships, connectedness, prediction, and control. It may also be true that quantum phenomena apply somewhat to us large-sized objects, literally more than we had thought. Our brain cells "are sensitive enough to register the absorption of a single photon . . . and thus sensitive enough to be influenced by the whole panoply of odd, quantum-level behavior" writes Zohar (1990, 79). And Wolf notes that "Instead of finding quantum mechanics restricted to ever tinier corners of the universe, we physicists are finding its applicability ever increasing to larger and larger neighborhoods of time and space" (1981, xiv).

Because the quantum world is so strange, its chroniclers reach for new metaphors. Zohar depicts it as "a vast porridge of being where nothing is fixed or measurable . . . somewhat ghostly and just beyond our grasp" (1990, 27). Capra sees it as "dynamic patterns continually changing into one another—the continuous dance of energy" (1983, 91). Others say that it is a place where "everything is interconnected like a vast network of interference patterns" (in Lincoln 1985, 34). In 1930, astronomer James Jeans created my own favorite image of this new world: "The universe begins to look more like a great thought than like a great machine" (in Capra 1983, 86).

When the world ceased to be a machine, when we began to recognize its dynamic qualities, many familiar aspects of it disappeared. In the work of quantum theorists, "things" disappeared. Although some scientists still conduct a determined search for the basic building blocks of matter, other physicists have abandoned this as a final, futile quest of reductionism. They gave up searching for things finite and discrete because, as they experimented to find elementary particles, they found "things" that changed form and properties as they responded to one another, and to the scientist observing them. "In place of the tiny billiard balls moved around by contact forces," Zohar writes, "there are what amount to so many patterns of active relationship, electrons and photons, mesons and nucleons that tease us with their elusive double lives as they are now position, now momentum, now particles, now waves, now mass, now energy—and all in response to each other and to the environment" (1990, 98).

In the quantum world, relationships are not just interesting; to many physicists, they are *all* there is to reality. One physicist, Henry Stapp, describes elementary particles as, "in essence, a set of relationships that reach outward to other things" (in Capra 1983, 81). Particles come into being ephemerally through interactions with other energy sources. We give names to each of these sources—physicists still identify neutrons, electrons, and other particles—but they are "intermediate states in a network of interactions." Physicists can plot the probability and results of these interactions, but *no particle can be drawn independent from the others*. What is important in any diagram is the overall process by which elements meet and change; analyzing them for more individual detail is simply not possible (Zukav 1979, 248–50). (See the illustration on page 35.)

In organizations, we are at the edge of this new world of relationships, wondering if the new charts are true, still fearing that if we follow them we will fall off into nothing. A mariner, perched high in the crow's nest, sometimes cries "Land ho" on faith. Knowing what to look for, knowing how hills appear on the horizon, knowing how to tell clouds from land—still, sometimes, the call is an act of faith. Sighting a world of quantum organizations requires such faith. But as we become more familiar with the quantum world, a few of its organizational features emerge from the fog, their outlines just discernible.

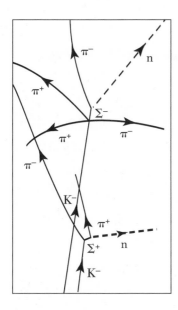

Particle Interaction—In a moment of time too brief to comprehend, K mesons enter a bubble chamber. As they interact with different energy potentials, twelve different particles appear temporarily. *Courtesy of the Lawrence Berkeley Laboratory, University of California.*

This world of relationships is rich and complex. Gregory Bateson (1980) speaks of "the pattern that connects" and urges that we stop teaching facts—the "things" of knowledge—and focus, instead, on relationships as the basis for all definitions. With relationships, we give up predictability and open up to potentials. Several years ago, I read that elementary particles were "bundles of potentiality." I began to think of all of us this way, for surely we are as undefinable, unanalyzable, and bundled with potential as anything in the universe. None of us exists independent of our relationships with others. Different settings and people evoke some qualities from us and leave others dormant. In each of these relationships, we are different, new in some way.

If nothing exists independent of its relationship with others, we can move away from our need to think in terms of separate, polar opposites. For years I had struggled conceptually with a question I thought important: In organizations, which is the more important influence on behavior—the system or the individual? The quantum world answered that question for me with a resounding "Both." There are no either/ors. There is no need to decide between two things, pretending they are separate. What is critical is the *relationship* created between two or more elements. Systems influence individuals, and individuals call forth systems. It is the relationship that evokes the present reality. Which potential becomes real depends on the people, the events, and the moment.

Prediction and replication are, therefore, impossible. While this is no doubt unsettling, it certainly makes for a more interesting world. People stop being predictable and become surprising. Each of us is a different person in different places. This doesn't make us inauthentic; it merely makes us quantum. Not only are we fuzzy; the whole universe is.

One source of universal fuzziness comes from the fact that elementary matter is inherently two-faced. It possesses two very different identities. Matter can show up as particles, specific points in space; or it can show up as waves, energy dispersed over a finite area. Matter's total identity (known as a wave packet) includes the potential for both forms—particles and waves. This is the Principle of Complementarity, and at heart, if I may give it a philosophical slant, it speaks of unity expressed as diversity.

However, these two complementary identities of one particle cannot be studied simultaneously as a unified whole. Here, we are thwarted by another major principle of quantum physics, Heisenberg's Uncertainty Principle. We can measure the particle aspect, or the wave aspect—either location or movement—but we can never measure *both* at the same time: "While we can measure wave properties, or particle properties, the exact properties of the *duality* must always elude any measurement we might hope to make. The most we can hope to know about any given wave packet is a fuzzy reading of its position and an equally fuzzy reading of its momentum" (Zohar 1990, 27). It is this "vast porridge of being" that sucks in like quicksand all our hopes for a deterministic, quantifiable universe.

These two principles ask us to fundamentally change our relationship to measurement and observation. If quantum matter develops a relationship with the observer and changes to meet his or her expectation, then how can there be scientific objectivity? If the scientist structures an experiment to study wave properties, matter behaves as a wave. If the experimenter wants to study particles, matter obliges and shows up in particle form. The act of observation causes the potentiality of the wave packet to "collapse" into one or the other aspect. One potential becomes realized while the others instantly disappear. Before the observer acts, an endless profusion of possibilities continues to be available. But once the observer chooses what to perceive, "the effect of perception is immediate and dramatic. All of the wave function representing the observed system collapses, except the one part, which actualizes into reality" (Zukav 1979, 79).

Several years ago, organizational theorist Karl Weick called attention to a similar observation dilemma in organizations, what he termed *enactment*. We participate, he noted, in the creation of our organizational realities: "The environment that the organization worries about is put there by the organization." Weick's observation, from a social science perspective, displays a sensibility quite similar to that of quantum physicists. There is no objective reality; the environment we experience does not exist "out there." It is co-created through our acts of observation, what we choose to notice and worry about. If we truly embraced this sensibility in our organizational life, we would no longer waste time arguing about the "objective" features of the environment. Conflicts about what's true and false would disappear in the exploration of multiple perceptions. Weick encourages us to move away from arguing about who's right and who's wrong, and instead to focus our concerns on issues of effectiveness, on reflective questions of what happened, and what actions might have served us better. We could stop arguing about truth and get on with figuring what works best (1979, 152, 168–69).

Weick also suggested a new perspective on organizational analysis. Acting should precede planning, he said, because it is only when we act to implement something that we create the environment. Until we begin this interaction with the environment, how can we formulate our thoughts and plans? In strategic planning, we act as though we are responding to a demand from the environment, but, in fact, Weick argued, we *create* the environment through our own intentions. Strategies should be "just-in-time . . . , supported by more investment in general knowledge, a large skill repertoire, the ability to do a quick study, trust in intuitions, and sophistication in cutting losses" (1979, 223, 229).

Weick's understanding of how systems and their environments co-create themselves has been further developed in recent years by a major rethinking of the entire field of strategic planning (see Mintzberg 1993). Many former planning advocates now speak about strategic *thinking* rather than planning. They emphasize that organizations require new skills. Instead of the ability to analyze and predict, we need to know how to stay acutely aware of what's happening now, and we need to be better, faster learners from what just happened. Agility and intelligence are required to respond to the incessant barrage of frequent, unplanned changes. Jack Welch, legendary CEO of General Electric, says that in this modern world of constant flux, "predicting is less important than reacting" (*USA Today*).

These shifts in how we think about strategy and planning are important to notice. They expose the fact that for many years and many dollars, we have invested in planning processes derived from Newtonian beliefs. How many companies made significant gains and consistent progress because of elaborate and costly strategic plans? Very few. A quantum perspective provides one powerful explanation for these failures. If there is no objective reality out there, then the environment and our future remain uncreated until we engage with the present. We must interact with the world in order to see what we might create. Through engagement in the moment, we evoke our futures.

This is not a suggestion that organizations exist in a totally reactive state. There is an essential role for *organizational intent and identity*. Without a clear sense of who they are, and what they are trying to accomplish, organizations get tossed and turned by shifts in their environment. No person or organization can be an effective co-creator with its environment without clarity about who it is intending to become (see Chapter Seven).

So many of the things in organizations that we argue and worry about come from our belief in objective reality. Something is out there, we believe, challenging our skills of analysis and perception. We just have to hire the right experts in order to see it clearly. But this search for discernible, objective futures has been, if we can admit it, a great cosmic joke. We thought we could nail down reality, get it in our sights, or maybe even line up our ducks, but how is that possible in this elusive world of potentials? We've been playing with "vast networks of interference patterns," with "the continuous dance of energy." The world is not an independently existing thing. It's a complex, never still, always weaving tapestry.

To live in a quantum world, to weave here and there with ease and grace, we need to change what we do. We need fewer descriptions of tasks and instead learn how to facilitate *process*. We need to become savvy about how to foster relationships, how to nurture growth and development. All of us need to become better at listening, conversing, respecting one another's uniqueness, because these are essential for strong relationships. The era of the rugged individual has been replaced by the era of the team player. But this is only the beginning. The quantum world has demolished the concept that we are unconnected individuals. More and more relationships are in store for us, out there in the vast web of life.

Even organizational power is purely relational. One evening, I had a long, exploratory talk with a wise friend who told me that "power in organizations is the capacity generated by relationships." It is an energy that comes into existence through relationships. Ever since that conversation, I have changed what I pay attention to in an organization. Now I look carefully at a workplace's capacity for healthy relationships. Not its organizational form in terms of tasks, functions, span of control, and hierarchies, but things more fundamental to strong relations. Do people know how to listen and speak to each other? To work well with diverse members? Do people have free access to one another throughout the organization? Are they trusted with open information? Do organizational values bring them together or keep them apart? Is collaboration truly honored? Can people speak truthfully to one another?

Because power is energy, it needs to flow through organizations; it cannot be bounded or designated to certain functions or levels. What gives power its charge, positive or negative, is the nature of the relationship. When power is shared in such workplace redesigns as participative management and self-managed teams, positive creative power abounds. For years, many people and researchers have described the positive impacts of these new relationships, power that shows up as significant increases in productivity and personal satisfaction (see Weisbord 1987, Daft and Lengel 1998).

In other workplaces, leaders attempt to force better results through coercion and competition; sometimes they exhibit a flagrant disregard for people and their abilities. In such organizations, a high level of energy is also created, but it's entirely negative. Power becomes a problem, not a capacity. People use their creativity to work *against* these leaders, or in spite of them; they refuse to contribute positively to the organization.

The learning for all of us seems clear. If power is the capacity generated by our relationships, then we need to be attending to the *quality* of those relationships. We would do well to ponder the realization that love is the most potent source of power.

The quantum world asks us to contemplate other mysteries as well. It reveals the webs of connection that are everywhere, and tantalizes us with a question: How do influence and change occur within a web? Physicists have observed a level of connectedness among seemingly separate particles, even if separated by huge distances. After 1930, a great debate raged among the premier physicists, especially between Niels Bohr and Albert Einstein. Could matter be affected by "non-local causes"? Could matter be changed by influences that travel faster than the speed of light? Einstein was so repelled at the idea of a universe where cause could happen at a distance that he designed a thought experiment with two other physicists (the EPR experiment) to disprove the whole idea.

His experiment launched a lively debate in physics, and thirty years later, with the debate still raging, physicist John Bell constructed a mathematical proof to show that "instantaneous action-at-a-distance" could occur in the universe. Finally, in 1982 (and subsequently established in many other experiments), French physicist Alain Aspect conducted actual physical experiments proving that elementary particles are, indeed, affected by connections that exist invisibly across time and space (Gribbin 1984, 227ff).

Here is one example of how action-at-a-distance is confirmed. Two electrons are first paired together or correlated. Tests are then conducted to determine whether such paired electrons, even when separated, will continue to act as one unified electron. Will their relationship survive at a distance? To determine whether these electrons behave as one, physicists can test their spin. Electrons spin along an axis, either up and down or side to side. However, being quantum phenomena, these axes do not preexist as objective reality. They exist only as potentials *until* the scientist decides on which axis to measure. There is no fixed spin to the electron; its spin appears based on what the scientist chooses to test for. The electrons respond to the scientist's choice of measures. (If this is hard to comprehend, remember that the quantum realm is weird even to scientists.) Once two electrons have been paired, if one is observed to spin up, the other will spin down, or if one is observed to spin right, the other will spin left.

In this experiment, the two paired electrons are separated. Theoretically, they could be across the universe. No matter the distance, at the moment one electron is measured for its spin—say that a vertical axis is chosen—the second electron will instantaneously display a vertical, but opposite, spin. How does this second electron, so far away, know which axis was chosen to measure?

Formerly, scientists believed that nothing travels faster than the speed of light, yet these experiments seem to disconfirm that. One explanation that physicists offer is that the two electrons are linked by non-visible connections; they are, in fact, an indivisible whole that cannot be broken into parts, even when separated by space. When an attempt is made to measure them as discrete parts, scientists get stymied by the fact of their invisible connectedness.

In our day-to-day search for order and prediction, we are driven crazy by non-local causality. In spite of the best plans, we experience influences that we can't see or test, and strange occurrences that pop up everywhere. We all have been forced to deal with unintended consequences of our well-intended plans. We thought we were doing something helpful to solve a problem, and suddenly we are confronted with eight new problems created by our initial solution. There is no way to prevent these troubling consequences. We can never do sufficient planning to avoid them, because we can't possibly see all the connections that are truly there. When we take a step or make a decision, we are tugging at webs of relationships that are seldom visible but always present.

We have broken the world into parts and fragments for so long now that we are ill-prepared to see that a different order is moving the whole. According to British physicist David Bohm, "The notion that all these fragments are separately existent is evidently an illusion, and this illusion cannot do other than lead to endless conflict and confusion" (1980, 1). I believe that one of our greatest challenges, after so many centuries of separation and fragmentation, is to discover new ways of thinking and sensing that allow us to comprehend the whole. This is still uncharted territory, and it requires the earnest explorations of many of us (see Chapter Eight).

At present, our most sophisticated way of acknowledging the world's complexity is to build elaborate system maps, which are most often influenced by a quest for predictability. When we create a map—displaying what we think are all the relevant elements and interactions—we hope to be able to manipulate the system for the outcomes we desire. We are thinking like good Newtonians. But what we hope for is not possible. There are no routes back to the safe harbor of prediction—no skilled mariners able to determine a precise course across the quantum ocean. The challenge for us is to see past the innumerable fragments to the whole, stepping back far enough to appreciate how things move and change as a coherent entity. We live in a very fuzzy world, where boundaries have an elusive nature and seldom mean what we expect them to mean. The illusory quality of these boundaries will continue to drive us crazy as long as we focus on trying to specify them in more detail, or to decipher clear lines of cause and effect between concepts that we treat as separate, but which aren't.

There are no familiar ways to think about the levels of interconnectedness that seem to characterize the quantum universe. Instead of a lonely void, with isolated particles moving through it, space appears filled with connections. This is why the metaphors turn to webs and weaving, or to the world as a great thought. Gravity is an everyday example of "action-at-a-distance," and scientists have created other "fields," unseen forces that organize space, to explain the connections they observe (see Chapter Three). The more provocative view, expressed in Bohm's work, is that at a level we can't discern, there is an unbroken wholeness. If we could look beneath the surface, we would observe an "implicate order" out of which seemingly discrete events arise (Bohm 1980).

Quantum leaps are an excellent example of quantum interconnectedness.

Technically, these leaps are abrupt and discontinuous changes, where an electron jumps from one orbit to another without passing through any intermediate stages. It's in one place and then suddenly it's in another, and there are no transition points en route to mark the journey. Physicists can calculate the probability of a jump occurring, but not precisely when it will take place. What is at work here is a whole system invisibly creating the conditions that suddenly enable it to jump to a new place. Because it is impossible to ever know everything about the whole, it is impossible to ever predict exactly where or when influences will manifest. This is hardly a comforting thought to those of us trying to lead organizations, yet the imagery of quantum leaps more accurately reflects my experience of organizational and societal change than any other.

I know of no better theory to explain the sudden fall of the Berlin Wall, for example. Before that event, there were many small changes going on throughout East Germany, most of which were not visible to anyone beyond their immediate neighborhood. But each small act of defiance or new way of behaving occurred within a whole fabric. Each small act was connected invisibly to all others. The global impact suddenly became visible in those few days when people tore the Wall down. The fall of the Berlin Wall demonstrates the power of "think globally, act locally." It proves that local actions can have enormous influence on a monstrous system that had resisted all other political attempts to change it. Germany could not be reunified by traditional power politics, or by high-level leaders from powerful nations. It was local actions within the system, combined with many other influences globally, that coalesced into a moment of profound change.

In a web, the potential impact of local actions bears no relationship to their size. When we choose to act locally, we may be wanting to influence the entire system. But we work where we are, with the system that we know, the one we can get our arms around. From a Newtonian perspective, our efforts often seem too small, and we doubt that our actions will make a difference. Or perhaps we hope that our small efforts will contribute incrementally to large-scale change. Step by step, system by system, we aspire to develop enough mass or force to alter the larger system.

But a quantum view explains the success of small efforts quite differently. Acting locally allows us to be inside the movement and flow of the system, participating in all those complex events occurring simultaneously. We are more likely to be sensitive to the dynamics of this system, and thus more effective. However, changes in small places also affect the global system, not through incrementalism, but because every small system participates in an unbroken wholeness. Activities in one part of the whole create effects that appear in distant places. Because of these unseen connections, there is potential value in working anywhere in the system. We never know how our small activities will affect others through the invisible fabric of our connectedness. I have learned that in this exquisitely connected world, it's never a question of "critical mass." It's always about *critical connections*.

Those who have used music metaphors to describe working together, especially jazz metaphors, are sensing the nature of this quantum world. This world demands that we be present together, and be willing to improvise. We agree on the melody, tempo, and key, and then we play. We listen carefully, we communicate constantly, and suddenly, there is music, possibilities beyond anything we imagined. The music comes from somewhere else, from a unified whole we have accessed among ourselves, a relationship that transcends our false sense of separateness. When the music appears, we can't help but be amazed and grateful.

My growing sensibility of this quantum world has profoundly affected my practice in organizations. Now I struggle to remain aware of the system as a system and to give up my well-trained abilities to reduce and separate things as the route to understanding. I concentrate much more on processes now, focusing on qualities rather than quantities, paying more attention to things like pattern, direction, feel, and the internal rhythm of what's happening. Long ago I gave up looking for straightforward cause and effect. I feel similarly that positioning things as polarities doesn't help—we need to stop drawing lines of opposition and try to understand the "and" of one and one.

I don't personally spend time anymore on elaborate plans or time lines. I want to use the time formerly spent on detailed planning and analysis to create the organizational conditions for people to set a clear intent, to agree on how they are going to work together, and then practice to become better observers, learners, and colleagues as they co-create with their environment. And I have learned that great things are possible when we increase participation. I always want more people, from more diverse functions and places, to be there. I am always surprised by what people can create as they explore the webs of relation and caring that connect them. Finally, I no longer argue about what is real. We each construct reality, and when I become curious about this, I learn a great deal from other people. I expect them to see things differently from me, to surprise me.

Underlying each of these changes in practice is a profound change in sensibility—I have given up trying to control anything. It has taken me a long while to learn this, but I finally understand that the universe refuses to cooperate with my desire to play God.

Sometimes I receive calls from consultant friends who are deep into a project and very frustrated. In one such call, a friend reported that his client organization had collected data, defined five key problem areas, and created task forces to solve each of those issues. Yet the managers were having problems coordinating the task forces. The longer the task forces studied the issues, the more they were seeing the problems as interrelated. Threads of interconnections were everywhere, yet the five groups were still acting separately from one another. The result was fatigue and impatience. People simply wanted to get on with implementing something; *anything* would be a relief after so many deadening meetings and detailed plans.

As I listened to my colleague, I shared his "Newtonian despair." I knew what he was feeling; I knew where things were headed if he continued to pursue these separate activities. We talked for some time about bringing the whole system together to access a deeper system's intelligence, but he was struggling to believe that this would help. He wanted to respond in new ways, but lacked a richer vision of what to do, of how to be in this world with greater trust. I wanted to be much more helpful, but in that moment, I failed him. I couldn't adequately convey the strangeness and beauty of this world, or help him believe enough in its inherent orderliness. These were things that I was only beginning to discover myself.

I felt as Heisenberg must have, when he walked those streets at dawn, begging for new insights into the universe. I, too, can feel the ground shaking. Many of us hear its deep rumblings. Any moment now, the earth will crack open and we will stare into its dark center. Into that smoking caldera, we will be asked to throw most of what we have treasured, most of the techniques and tools that have made us feel competent. We know what we must do. And when we finally step forward to do it, when we have made our sacrificial offerings to the gods of understanding, then the ruptures will cease. Healing waters will cover the land, giving birth to new life, burying forever the ancient, rusting machines of our past understandings. And on these waters we will set sail to places we now can only imagine. There we will be blessed with new visions and new magic. We will feel once again like creative participants in this mysterious world. But for now, we wait. An act of faith. Land ho.

Although we know a great deal about the way fields affect the world as we perceive it, the truth is no one really knows what a field is. The closest we can come to describing what they are is to say that they are spatial structures in the fabric of space itself.

—Michael Talbot

Chapter 3

Space Is Not Empty:
Invisible Fields That Shape Behavior

I n Utah, the sky is everywhere—blue, open, insistent on attention. It soars over mountains and dives into long valleys, showing off its crystal clarity. At night, it is even more an exhibitionist. A friend, after a long flight from Hartford, sat rocking on my lawn swing far past midnight, tired, yawning, but unable to move. The stars would not let her go. For me, moving here—and living with these stars and sky—has been an experience in space. I have felt myself expanding into this vastness, felt my boundaries open, my vision lift, my internal defenses dissolve. With so much space, there is no place to go but out.

Space is the basic ingredient of the universe; there is more of it than anything else. Even at the microscopic level of atoms, where we would expect things to be dense and compact, there is mostly space. Within atoms, subatomic particles are separated by vast distances, so much so that an atom is 99.99 percent empty. Everything we touch, including our bodies, is composed of these empty atoms. We are far more porous than our dense bodies indicate. In fact, we are as void, proportionately, as intergalactic space (Chopra 1989, 96).

In Newton's universe, the emptiness of space created a sense of unspeakable loneliness. Matter, alone and isolated, moved bravely through the void, making a solo journey, meeting others rarely, traveling always across wide gulfs that stretched on to infinity. This lonely universe has, for a long time, affected our self-expression in all ways, from existentialist philosophies that say there is no meaning to our lives except that which we create for ourselves, to the heroic individuals of American history, lonely champions (both Western and corporate) who succeeded in spite of great odds. It was difficult to effect change in such an vast, lonely world. It required generating energy of sufficient strength to propel oneself through space, enduring long enough to reach another object and cause it to respond. Newton's world of cause and effect, of force acting upon force, required great expenditures of personal energy to get someone else moving, vast regions of space to traverse to get something done. Not only did it feel lonely; it was exhausting.

Something strange has happened to space in the quantum world. No longer is it a lonely void. Space everywhere is now thought to be filled with fields, invisible, non-material influences that are the basic substance of the universe. We cannot see these fields, but we do observe their effects. They have become a useful way to explain action-at-a-distance, a descriptor for how change occurs without the direct exertion of one element needing to shove another into place.

In scientific thought, field theory developed in several different areas years before quantum physics as an attempt to explain action-at-a-distance. (The word *field* was taken from the name given to the background on heraldic shields.) Newton introduced the first field, gravitation. In his model, gravity originated from a center of force, such as the earth, and spread out from there into space. Imaginary lines of force filled space, attracting objects toward the earth. In Newton's model of gravitational pull, a force emanated from one source, acting on another.

Einstein developed a different view of the gravitational field. In relativity theory, gravity acts to structure space. The reason objects are drawn to earth is because space-time curves in response to matter. Rather than a force, gravity is understood as a medium, the invisible geometry of space.

In our day-to-day lives, we have direct experience with other fields besides gravity. Just place iron filings near a magnet. The specific patterns that form around the magnet are due to the invisible magnetic field. We also experience the effects of fields every time we turn on a light or plug in an appliance. Our modern electrical generating stations spin huge magnets, creating magnetic fields that then create electrical fields, which send out currents of electrons.

Fields are conceived of in many different ways. The gravitational field is thought to be a curved structure in space-time; electromagnetic fields create disturbances that manifest as electromagnetic radiations; quantum fields, perhaps a different field for each particle, are energy, manifesting into form when two fields intersect. But in all of these theories, fields are unseen forces, invisible influences in space that become apparent through their effects (Wilczek and Devine 1988, 155–64; Zukav 1979, 199–200).

Early advances in field theory came about because nineteenth-century scientists such as Michael Faraday and James Maxwell chose to concentrate not on specific particles, but on space. Intuitively, they sensed that space was not empty but instead was, in a modern physicist's phrase, "a cornucopia of invisible but powerfully effective structure" (Wilczek and Devine 1988, 156). Faraday and Maxwell made a conscious shift in vision, as we do when we look from close to distant objects, and in that shift they led the way into a universe of busy, bustling space. It was an important shift in focus—to look behind the small, discrete, visible structures to an invisible world filled with mediums of connection. (See Aurora Borealis photo, created by electromagnetic and energy fields in the atmosphere, in the color section.)

Frank Wilczek and Betsy Devine, he a physicist, she an engineer turned writer, created an effective image for thinking about these invisible powers that exert visible influence. If we were to observe fish, unaware of the medium of water in which they swim, we would probably look for explanations of their movements in terms of one fish influencing another. If one fish swam by and we observed the second fish swerving a little, we might think that the first fish was exerting a force on the second. But if we observed all the fish deflecting in a regular pattern, we might begin to suspect that some other medium was influencing their movements. We could test for this medium, even if it were still invisible to us, by creating disturbances in it and noting the reactions of the fish. The space that is everywhere, from inside atoms to the cosmos, is more like this ocean, filled with fields that exert influence and bring matter into form.

The world described by new science is teasing and enticing in many ways. Fields fit right in. Biologist Rupert Sheldrake describes them as "invisible, intangible, inaudible, tasteless and odorless" (1995, 1988). They are unapproachable through our five senses, yet in quantum theory, they are as real as particles. Writer Gary Zukav terms them the substance of the universe. The things we see or observe in experiments, the physical manifestations of matter as particles, are a secondary effect of fields. Particles may come into existence, temporarily and briefly, when two fields intersect. At the point of meeting, where their energies interact, particles appear. The fact that particles appear and disappear like quick-change artists is a result of continual interactions between different fields. Although we have thought of particles as the basic building blocks of matter, in fact they are transitory, just brief moments of meeting recorded as observable matter. This leads to a puzzling situation. Physical reality is not only physical. Fields are considered real, but they are non-material.

This paradox pushes us into important new territory, urging us further away from our "thing" thinking, away from a universe of parts linked together tenuously. Fields encourage us to think of a universe that more closely resembles an ocean, filled with interpenetrating influences and invisible forces that connect. This is a much richer portrait of the universe; in the field world, there are potentials for influence everywhere, whenever two energies meet: "The Newtonian picture of a world populated by many, many particles, each with an independent existence, has been replaced by the field picture of a world permeated with a few active media. We live amid many interpenetrating fields—each filling space. The laws of motion, in field language, are rules for flows in this ocean. And the rules of transformation are, in this picture, telling us what . . . reactions occur among the components of the universal ocean" (Wilczek and Devine 1988, 163).

In biology, Sheldrake has created a controversial concept of fields. He has postulated the existence of morphic fields that influence the behavior of species. This type of field possesses very little energy of its own, but it is able to shape energy that comes from another source. Morphic fields are built up through the skills that accumulate as members of the same species learn something new (Sheldrake 1995, 82). After some number (not specified) of a species have learned a behavior, such as bicycle riding, others of that same species will be able to learn that skill more easily. The behavior collects in the morphic field, and when an individual's energy combines with it, the field patterns the behavior of that individual. They don't have to actually learn the skill; they pull it from the field. They learn it through "morphic resonance," a process Sheldrake describes as individuals being influenced by others like them. These fields, says Bohm, provide "a quality of form that can be taken up by the energy of the receiver" (in Talbot 1986, 68; see also Sheldrake 1988, 1995).

The imagery provided in any of these field theories is quite provocative, because it invites us to contemplate space differently. We already live and work with a new awareness of space. Through electronic networks, we reach into the invisible to feed ourselves with information. We rely on information to move through the ethers, retrievable from who-knows-where. But who has seen *cyberspace?* The invisible is more of an active player in our lives than ever before.

But it is time to think beyond cyberspace to what else might be going on in the space of our organizations. It might be that our communal space is filled with these "interpenetrating influences and invisible forces that connect." How would we discern organizational fields? If we understood more about them, could they assist us in creating the behaviors we desire? It seems important to at least contemplate that something might be going on in the spaces among us. Space is not empty. Unseen influences affect behavior.

For several years now, leaders have been encouraged to consider the impact of non-material forces in organizations—culture, values, vision, ethics. Each of these concepts describes a *quality* of organizational life that can be observed in behavior yet doesn't exist anywhere independent of those behaviors. Once when I was working on customer service for a large retail chain, I asked employees to visit several stores. After spending time in many stores, we all compared notes. To a person, we agreed that we could "feel" good customer service when we walked in a store. We tried to get more specific by looking for visual cues, merchandise layouts, facial expressions—but none of these were sufficient to explain the sure sense we had when we walked into that store that we would be treated well. Something else was going on. We could feel it; we just couldn't describe *why* we felt it.

It seems to me that field theory provides a plausible explanation to this and other organizational mysteries. Thinking about the possibility of organizational fields is an interesting exercise in metaphoric thinking. It can help us contemplate our experience with unseen influences, and with behaviors that may have been difficult to change through more direct approaches. What is it that influences employee behavior or that encourages employees to practice things like excellent customer service? This is where field theory can lead us to new questions. We could ask about the messages that fill the space of the organization, thinking of these messages as an organizational field that is influencing behavior. We would look to discern what's in the field, whether messages there are congruent or discordant. We might discover that while we say we want outstanding customer service, there are other messages that exert reverse pressures. Perhaps people are being signaled that they must make their quotas this quarter no matter what. Or that they must make their boss look good above all other considerations.

We can never see a field, but we can easily see its influence by looking at behavior. To learn what's in the field, look at what people are doing. They have picked up the messages, discerned what is truly valued, and then shaped their behavior accordingly. When organizational space is filled with divergent messages, when only contradictions float through the ethers, this invisible incongruity becomes visible as troubling behaviors. Because there is no agreement, there are more arguments, more competition, more power plays. People say one thing and mean another. Nobody trusts anybody. The organization changes direction frequently and can't find its way.

While I have no need to affirm the actual presence of fields in those retail stores I visited years ago, I am positive that in each one where customers felt welcome, there was a leader who, in word and deed, filled space with clear and consistent messages about how customers were to be served. The field was strong in its congruence; it influenced behavior only in one direction. Because of the power of this field, the outcome was assured: outstanding customer service.

The invisible influences that field theory exposes can help us manage other amorphous aspects of organizational life. For example, *vision*—organizational clarity about purpose and direction—is a wonderful candidate for field theory. In linear fashion, we have most often conceived of vision as designing the future, creating a *destination* for the organization. We have believed that the clearer the image of the destination, the more force the future would exert on the present, pulling us to that desired state. It's a very strong Newtonian image, much like the old view of gravity. But what if we changed the science and looked at vision as a field?

If vision *is* a field, think about what we could do differently to use its formative influence. We would start by recognizing that in creating a vision, we are creating a power, not a place, an influence, not a destination. This field metaphor would help us understand that we need congruency in the air, visionary messages matched by visionary behaviors. We also would know that vision must permeate through the entire organization as a vital influence on the behavior of all employees. And we would feel genuinely threatened by incongruous acts because we would understand their disintegrating effects on what we dream to accomplish. We would become an organization of integrity, where our words would be seen and not just heard.

Several years ago, a garbage-can metaphor was introduced into our thinking about organizations. It created a provocative view of organizational "space" as a continual mixture of people, solutions, choices, and problems circulating aimlessly, every so often coinciding and creating a decision at that juncture: "An organization is a collection of choices looking for problems, issues and feelings looking for decision situations in which they might be aired, solutions looking for issues to which they might be the answer, and decision makers looking for work" (Cohen, March, and Olsen 1974).

This is a cynical but real view of life in a Newtonian organization, discrete pieces wandering about, colliding or avoiding collision, veering off in unexpected directions—organizational anarchy relieved by occasional moments of accidental coherence. This metaphor is still a harrowing but familiar description of the irrational energies that stalk the halls of too many organizations. The task of creating order in a garbage can, of imposing structure and meaning on a smelly mess, is virtually impossible.

But with a quantum sensibility, there are new possibilities for how to create order. Organizational behavior is influenced by the invisible. If we attend to the fields we create, if we help them shine clear with coherence, then we can clean up some of the waste of organizational life.

In many ways, we already know what powerful organizers fields can be. We have moved deeper into understanding these invisible allies with the recent focus on organizational culture, values, and purpose. We see that these are important, even when we don't quite know why. Robert Haas, former CEO of

Levi Strauss & Co., calls these the "conceptual controls. . . . It's the ideas of a business that are controlling, not some manager with authority" (in Howard 1990, 134). If we understand ideas as real forces in the organization, as fields, I believe we have a better image for understanding why concepts control as well as they do. But the shift in imagery changes the nature of our attention.

In a field view of organizations, we attend first to clarity. We must say what we mean and seek for a much deeper level of integrity in our words and acts than ever before. And then we must make certain that everyone has access to this field, that the information is available everywhere. Vision statements move off the walls and into the corridors, seeking out every employee, every recess in the organization. In the past, we may have thought of ourselves as skilled designers of organizations, assembling the pieces, drawing the boxes, exerting energy to painstakingly create all the necessary links, motivation, and structures. Now we need to imagine ourselves as beacon towers of information, standing tall in the integrity of what we say, pulsing out congruent messages everywhere. We need all of us out there, stating, clarifying, reflecting, modeling, filling all of space with the messages we care about. If we do that, a powerful field develops—and with it, the wondrous capacity to organize into coherent, capable form.

Let us remember that space is never empty. If it is filled with harmonious voices, a song arises that is strong and potent. If it is filled with conflict, the dissonance drives us away and we don't want to be there. When we pretend that it doesn't matter whether there is harmony, when we believe we don't have to "walk our talk," we lose far more than personal integrity. We lose the partnership of a field-rich space that can help bring order to our lives.

There is an irony here. Those who try to convince us to lead from values and vision, rather than from traditional forms of authority, don't seem to have enough substance. Their advice seems devoid of the structure and management controls that ensure order. Values, vision, ethics—these are too soft, too ethereal, to serve as management tools. How can they create the kind of order we need in the face of chaos? Newton's world justified those fears because it was a world with no internal coherence. Individual pieces spun off wildly on their individual trajectories. But if we look past Newton, if we change our field of vision, we see a world of more subtle ordering processes.

What if we slip out quietly along the curvature of space, out into its far reaches? What if, once there, we adjust our eyes to the invisible? There, instead of emptiness, we will see a richness of organizing energies. We once were made secure by things visible, by structures we could see. Now it is time to embrace the invisible. In a world where matter can be immaterial, where influences move among us unseen, why not contemplate the influence of fields? For such a little act of faith, space awaits, filled with possibilities.

Penetrating so many secrets,
we cease to believe in the unknowable.
But there it sits nevertheless
calmly licking its chops.

—H. L. Mencken

Chapter 4
The Participative Nature of the Universe

Schroedinger's cat is a classic thought problem in quantum physics. Physicist Erwin Schroedinger constructed the problem in 1935 to illustrate that in the quantum world nothing is real. We cannot know what is happening to something if we are not looking at it, and, stranger yet, nothing *does* happen to it until we observe it. Central to the quantum world, Zohar wrote, is the idea that "unobserved quantum phenomena are radically different from observed ones" (1990, 41).

The problem of the cat has not yet been resolved, but here is the thought experiment. A live cat is placed in a box. The box has solid walls, so no one outside the box can see into it. This is a crucial factor, since the thought experiment explores the role of the observer in evoking reality. Inside the box, a device will trigger the release of either poison or food; the probability of either occurrence is 50/50. Time passes. The trigger goes off, unobserved. The cat meets its fate.

Or does it? Just as an electron is *both* a wave and a particle until our observation causes it to collapse as *either* a particle or wave, Schroedinger argues that the cat is both alive *and* dead until the moment we observe it. Inside the box, when no one is watching, the cat exists only as a probability wave. It is possible to calculate mathematically (as a Schroedinger wave function) all of the cat's possible states. But it is impossible to say that the cat is living or dead until we observe it. It is the act of observation that determines the collapse of the cat's wave function and makes it either dead or alive. Before we peer in, the cat exists as probabilities. Our curiosity kills the cat. Or brings it back.

I have *never* understood the quantum logic of Schroedinger's cat, but I have let the problem ramble aimlessly in my mind, content to not engage with its counterintuitive nature. Yet just like a wave function, the possibilities of this idea grew unobserved until one day, in true quantum fashion, they "popped" and I had a moment of concrete recognition. I realized I had been living in a Schroedinger's cat world in every organization I had ever been in. Each of these organizations had myriad boxes, drawn in endless renderings of organizational charts. Within each of those boxes lay a "cat," a human being, rich in potential, whose fate was determined, always and irrevocably, by the act of observation.

It is common to speak of self-fulfilling prophecies and the impact these have on people's behavior. If a manager is told that a new trainee is particularly gifted, that manager will see genius emerging from the trainee's mouth even in obscure statements. But if the manager is told that his or her new hire is a bit slow on the uptake, the manager will interpret a brilliant idea as a sure sign of sloppy thinking or obfuscation. From studies on the impact of opportunity in organizations (Kanter 1977), we know that the "anointed" in organizations, those high flyers who move quickly through the ranks, are given at least some of their wings through our desire to observe them as winners. We endow their ideas and words with more credibility. We entrust them with more resources and better assignments. We have already decided that they will succeed, so we continually observe them with the expectation that they will confirm our beliefs.

Others in organizations go unobserved, forever invisible, bundles of potential that no one bothers to look at. Or they receive summary glances, are observed to be "dead," and are thereafter locked into jobs that provide them with no opportunity to display any new potential. In the quantum world, what you see is what you get. In human organizations, we play with Schroedinger's cat daily, determining the fate of all of us—our quality of aliveness or deadness—by what we decide to observe in one another. So it is not only quantum physicists who have to deal with the enigmas of observation. The observation problem is as real for us as it is for them.

In quantum physics, the observation problem has led scientists to develop various schools of thought, each focused on the role played by awareness. Is it awareness that evokes the world? Is there any such thing as reality independent of our acts of observation? Such questions touch upon ancient philosophical as well as scientific questions. Science writer and physicist Fred Allen Wolf asks: "If the world exists and is not objectively solid and preexisting before I come on the scene, then what is it? The best answer seems to be that the world is only a potential and not present without me or you to observe it. It is, in essence, a ghost world that pops into solid existence each time one of us observes it. All of the world's many events are potentially present, able to be but not actually seen or felt until one of us sees or feels."

These questions arise not because of the physicists' interest in philosophy, but because the issues emerge in actual quantum experiments. The double-slit experiment is the most frequently explained experiment that illustrates, among other things, the role of observation in the quantum world.

Most simply, this experiment involves electrons (or any other elementary particles) that must pass through one of two openings (slits) in a surface. After passing through one of these slits, each electron lands on a second surface, where its landing is recorded. A single electron passes through only one of the openings, but how it displays itself on the landing surface is affected by whether one or both slits are open at the time it passes through either one of them.

The electron, like all quantum entities, has two forms of being; it is both a wave and a particle. If both slits are open, the single electron acts as a wave, creating a pattern on the recording screen typical of the diffusion caused by a wave. If only one slit is open, the resulting pattern is that of discrete points, or the behavior of a particle.

On its way through one slit, the electron acts in a way that indicates it "knows" whether or not the second hole is open. It knows what the scientist is observing for and adjusts its behavior accordingly. If the observer tries to "fool" the subject by opening and shutting slits as the electron approaches the wall, the electron behaves in the manner appropriate for the state of the holes *at the moment* it passes through one. (For a detailed explanation of this experiment, see Gribbin 1984, 169–74.) The electron also knows if the observer is watching. If the recording apparatus is not on, the electron behaves differently than if it is being recorded. When the electron is not being observed, it exists only as a probability wave; unless someone is watching, "nature herself does not know which hole the electron is going through" (Gribbin 1984, 171).

Because nothing in the double-slit experiment can be explained by classical physics (or makes *any* sense to us laypersons), Physicist Richard Feynmann dubs these experiments "the only mystery," that which contains all of "the basic peculiarities of quantum mechanics" (in Gribbin 1984, 164). As non-physicists, we may think we have an easier time with the mysteries of such things as observation and the role of the observer, but it seems to me we would do well to linger longer with these quandaries, to explore how our perceptions of people and events shape the reality that we then end up struggling with so much.

Schroedinger's cat and the problem of observation pad quietly around our organizations in many forms. Fred Wolf says that "knowing is disrupting." Every time we go to measure something, we interfere. A quantum wave function builds and builds in possibilities until the moment of measurement, when its future collapses into only one aspect. Which aspect of that wave function comes forth is largely determined by *what* we decide to measure.

The physicist John Archibald Wheeler has been an eloquent proponent of the participative universe, a place where the act of looking for certain information evokes the information we went looking for—and simultaneously eliminates our opportunity to observe other information. For Wheeler, the entire universe is a participatory process, where we create not only the present with our observations, but the past as well. It is the existence of observers who notice what is going on that imparts reality to everything (Gribbin 1984, 212). When we choose to experiment for one aspect, we lose our ability to see any others. Every act of measurement loses more information than it gains, closing the box irretrievably and forever on other potentials.

The difficulties of observation raised by quantum sensibilities are problematic for all scientific inquiry, not just quantum physics. Modern science attempts to systematically observe the world around us. But science is not done in an objective world, free of observer influence. Every observation is preceded by a choice about *what* to observe (see Rose 1997, Ch. 2; Merchant 1980). No one, not scientists nor leaders nor children, simply observes the world and takes in what it offers. We all construct the world through lenses of our own making and use these to filter and select. We each actively participate in creating our worlds. Observation, then, is a very complex and important issue. "Whatever we call reality," Prigogine and Stengers advise, "it is revealed to us only through an active construction in which we participate" (1984, 293).

For leaders, being alert to the observation dilemma is critically important. Management is addicted to numbers, taking frequent pulses of the organization in surveys, monthly progress checks, quarterly reports, yearly evaluations. It is important to stay aware to the realization that no form of measurement is neutral. Every act of measurement loses more information than it gains. So how can we ensure that we obtain sound information to make intelligent decisions? How can we know what is the right information to look for? How can we remain open to the information we lost when we went looking for the information we got?

We don't often allow these questions to surface in organizations. We tend to focus on a few key indicators, or the opinions of those we trust. We worry more about the accuracy of the small bits of information we have and how best to analyze them than about the huge amounts of information we lose. Even when we attempt to look for data that are new and different, we still act as though that data exists "out there" and that we just have to find the appropriate lens or expert to get it. We still believe in objectivity, in truth, in hard data, in firm numbers. We have avoided coming to terms with the murky, fuzzy world that the observation dilemma exposes. As Fred Wolf said, "According to the quantum rules, we cannot ever know and experience simultaneously all that is in principle knowable. . . . One thing is clear, though: self plays a role in what is seen to be not-self" (1981, 80–81).

Yet how can we exist without objective information? How can we develop the information we need to do our work if we construct the worlds we inhabit? Just as the problem originates from the participative nature of the universe, so does the solution. Participation, seriously done, is a way out from the uncertainties and ghostly qualities of this nonobjective world we live in. We need a constantly expanding array of data, views, and interpretations if we are to make wise sense of the world. We need to include more and more eyes. We need to be constantly asking: "Who else should be here? Who else should be looking at this?"

Let me develop a quantum interpretation as to why participation is such an effective organizational strategy. In the traditional model, we leave the interpretation of data to senior or expert people. A few people, charged with interpreting the data, observe only a very few of the potentialities contained within that data. How often do we even think about all the data that goes unnoticed because we rely on these solitary observations?

Think of organizational data for a metaphoric moment as a quantum wave function, moving through space, rich in potential interpretations. If this wave of potentials meets up with only one observer, it collapses into only one interpretation, responding to the expectations of that particular person. All other potentials disappear from view and are lost by that solo act of observation. This one interpretation is then passed down to others in the organization. Most often the interpretation is presented as objective, which it is not, and definitive, which is impossible.

Consider how different it is, in quantum imagery, when data is recognized as a wave, rich in potential interpretations, and completely dependent on observers to evoke different meanings. If such data is free to move, it will meet up with many diverse observers. As each observer interacts with the data, he or she develops their own interpretation. We can expect these interpretations to be different, because people are. Instead of losing so many of the potentials contained in the data, multiple observers elicit multiple and varying responses, giving a genuine richness to the observations. An organization rich with many interpretations develops a wiser sense of what is going on and what needs to be done. Such organizations become more intelligent.

It would seem that the more participants we engage in this participative universe, the more we can access its potentials and the wiser we can become. We banish the ghosts of this ghostly universe by engaging in a different pattern of behavior—one in which more and more of us are included in the process of observing what is going on, and contributing our unique interpretations to the organization.

The truly miraculous organizational events I have participated in over the past several years are change efforts where the whole system is involved. As many as several hundred people are invited from all parts of the organization, including external stakeholders. For two to three days, they work intensely together to create shared visions of the organization's past, present, and future. The richness of the interpretations and the future scenarios they create have convinced me of the powers of participation. In these conferences, entirely new and surprising interpretations become available because the whole system is in the room, generating information, reflecting on itself and who it wants to become (see Weisbord and Janoff 1995). The miraculous enters in as the diversity of the group coalesces into a complex but unified vision of what they want to create together. This future vision is always far more powerful and ingenious than any individual could have possibly imagined.

The participative universe we inhabit has also deepened my understanding of the importance of "ownership," a term used to describe not only literal owners, but more importantly, the emotional investment of employees in their work. Ownership describes personal connections to the organization, the powerful emotions of belonging that inspire people to contribute. A tried and true maxim of my field of organizational behavior is that "people support what they create." Though I have preached, like many consultants before me, the values of psychological ownership, I now see that the quantum universe supports this concept even more strongly and explains *how* it creates real and tangible sources of energy.

We know that the best way to create ownership is to have those responsible for implementation develop the plan for themselves. No one is successful if they merely present a plan in finished form to others. It doesn't matter how brilliant or correct the plan is—it simply doesn't work to ask people to sign on when they haven't been involved in the planning process.

This is where the observation phenomenon of quantum physics has something to teach us. In quantum logic, it is impossible to expect any plan or idea to be real to people if they do not have the opportunity to personally interact with it. Reality is co-created by our process of observation, from decisions we the observers make about what we choose to notice. It does not exist independent of those activities. Therefore, we cannot *talk* people into our version of reality because truly nothing is real for them if they haven't created it. People can only experience a proposed plan by interacting with it, by evoking its possibilities through their personal processes of observation.

Think about what happens in your experience when you want to get something accepted. I see it all the time in meetings where a plan is being proposed. Even if it is excellent, it will be a long meeting in which the plan will be dissected, criticized, thrown out, brought back, and finally, almost always, approved in its initial form with only a few slight modifications. All of those participants, like the best scientists, need to observe the plan in detail, exploring its edges, searching out its interior, playing with its potentials. Each observer is evoking his or her version of the plan by the act of observation. After a period of sometimes maddening dissension, the dissections cease and people sit back content, filled with energy and commitment. Usually we endure these processes wondering why we have to go through them, especially because so often the agreed-upon plan bears a striking resemblance to what was proposed initially. But it is the *participation process* that makes the plan come alive as a personal reality. People can commit themselves because it has become real for them.

Participation, ownership, subjective data—each of these organizational insights that I gain from quantum physics quickly returns me to a central truth. We live in a universe where relationships are primary. Nothing happens in the quantum world without something encountering something else. Nothing exists independent of its relationships. We are constantly creating the world— evoking it from many potentials—as we participate in all its many interactions. This is a world of process, the process of connecting, where "things" come into temporary existence because of relationship.

Physicists have had a head start in becoming oriented to this new world of process. They pay attention to events and interactions rather than to things, thus becoming—in Gary Zukav's extended metaphor of the Wu Li Masters— observers of the dance (1979). But for us—as we sit in offices, structured into rigid relationships, besieged with stacks of data that accumulate daily, armed with complex formulas of interpretation—we have a long way to go before we can move onto that dance floor. It seems too strange, this realization that we participate in the creation of everything we observe.

It makes me wonder how we will design our organizations in the future. As we struggle with the designs that will replace bureaucracy, we must invent organizations where process is allowed its varied-tempo dance, where structures come and go as they support the work that needs to get done, and where forms arise to support the necessary relationships.

Physicists struggle with a similar dilemma when they try to diagram reactions between "things" that are not things until they are engaged with one another. There have been different ways of drawing the reactions by which particles appear, change, and participate in the creation of other particles. In two examples, lines converge from different points, forming new lines that go off in other directions. The elaborate lattice design of these drawings reinforces the idea that particles are best understood not as objects, but as occurrences, as temporary states in a network of reactions that go on and on.

Without understanding the physics in detail, I have been intrigued by some of the concepts in scattering-matrix diagrams (known as S-matrix diagrams).

S-Matrix Diagram—Particles come into being as intermediate states in a network of interactions. The energy of any particle can combine with other energy sources to create new particles. The lines represent particles as "reaction channels" through which energy flows. The circle denotes the area of interaction.

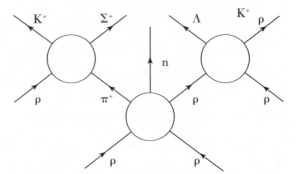

These diagrams represent a way of modeling the dynamic lives of high-energy particles and how they manifest into several different forms, depending on the energy available. I have spent hours staring at them, knowing they have something to teach me about organizational structure and how we might chart roles and relationships differently (see Capra 1976, 1983; and Zukav 1979).

The first thing that intrigues me about these diagrams is the concept of "reaction channels." In the diagrams, lines converge into a collision circle, from which other lines emerge. Each of these lines has a particle name attached to it, but the lines are best understood not as particles, not as things, but as "reaction channels," places where energy takes temporary form. Several different forms (particles) can emerge inside the reaction channels, depending on the amount of energy that is generated in the interactions.

Traditional organization charts are filled with lines connecting well-bounded boxes. It would be a breakthrough to think of the lines as reaction channels, places where energy meets up with other energy to create new possibilities. But S-matrices stretch my thinking even more because they demand that I stop thinking of roles or people as fixed entities. They lead me into the world of "no-things," where who you are depends on who you meet.

A subatomic particle is defined by its energy and by the network of relationships in which it exchanges energy. These subatomic particles, in Capra's words, "are not separate entities but interrelated energy patterns in an ongoing dynamic process. These patterns do not 'contain' one another but rather 'involve' one another. . ." (1983, 94). These particles are described as a *tendency* to participate in various reactions, a definition that honors the dynamic qualities of their existence. With S-matrix diagrams, physicists describe the processes of continual transformation, of emergence, decay, and the new forms that characterize high-energy particles. The result is an intriguing network of interactions, a structure of processes and potential relationships.

If I apply this to the roles and relationships described by organization charts, I get led to some different ideas about how to support good work. Roles mean nothing without understanding the network of relationships and the resources that are required to support the work of that person. In this relational world, it is foolish to think we can define any person solely in terms of isolated tasks and accountabilities. We need to be able to conceptualize the pattern of energy flows required for that person to do the job. We need to see any person's role as the place where energies meet to make something happen. The puzzling particle-interaction diagrams offer very different perspectives on what we must do to support individuals and their networks of relationships to work at transformative levels.

Unlike traditional organizational charts, S-matrix diagrams can also be rotated, thereby altering the reactions among the particle players. No one particle is the basic element or causative agent. Each has the capacity to interact with another and produce different outcomes. Rotating the diagrams changes the roles played by the different energies; what was a force influencing a reaction can, by turning the diagram, become a reaction channel influenced by other forces. Hierarchy and defined power are not what is important; what's critical is the availability of places for the exchange of energy.

Is it possible to think about organizational roles in this way, as focal points for interactions and energy exchanges? To any role's specific tasks and accountabilities, we would also consider how that role contributes energy to others. We would emphasize the interactions we needed, and we would want to ensure that the entire organization was capable of facilitating energy flows. Our attention would be directed to the energy and the relationships required to achieve a desired outcome. If we succeeded in thinking about organizations in this way, we could begin to create organizations of process and relationships, quantum organizations that worked more effectively in this relational universe.

Heisenberg describes the world of modern physics as one divided not "into different groups of objects but into different groups of connections." What is distinguishable and important, he says, are the kinds of connections. This is the world in which we must design and manage organizations, and there undoubtedly will be many more images from physics to challenge our prevailing ideas about organization.

Perhaps these are just the ramblings of one whose mind has gone fuzzy (like all quantum phenomena) from trying to understand quantum physics. But there *is* an urgent challenge to create organizations that respond to this new world of relationships in which we act as grand evocateurs of reality. Our old views constrain us. They deprive us from engaging fully with this universe of potentials.

When I think of all those wave functions filling space, rich in potentials, accumulating more and more possibilities as they fan out, I wonder why we limit ourselves so quickly to one idea or one structure or one perception, or to the idea that "truth" exists in objective form. Why would we stay locked in our belief that there is one right way to do something, or one correct interpretation to a situation, when the universe demands diversity and thrives on a plurality of meaning? Why would we avoid participation and worry only about its risks, when we need more and more eyes to be wise? Why would we resist the powerful visions and futures that emerge when we come together to co-create the world? Why would we ever choose rigidity or predictability when we have been invited to be part of the generative dance of life?

And why would we ever peer into that box expecting a dead cat, when just by our powers of observation we could bring that cat to life?

She who wants to have right without wrong,
Order without disorder,
Does not understand the principles
Of heaven and earth.
She does not know how
Things hang together.

—Chuang Tzu, fourth century B.C.

Chapter 5

Change, Stability, and Renewal:
The Paradoxes of Self-Organizing Systems

One day when a child, I stood beneath a swing frame that towered above me. Another child, older than me, told me of the time a girl had swung and swung until, finally, she looped over the top. I listened in silent awe. She had done what we only dreamed of doing, swung so uncontrollably high that finally not even gravity could hold her.

I think of this apocryphal story as I sit now in a small playground, watching my youngest son run from one activity to another. He has climbed, swung, and jumped, whirled around on a spinning platform, and wobbled along a rolling log until, laughing, he loses his balance. Now he is perched on a teeter-totter, waiting to be bumped high in the air when his partner crashes to the ground. Everywhere I look, there are bodies in motion, energies in search of adventure.

It seems that the very experiences these children seek are ones we avoid: disequilibrium, novelty, loss of control, surprise. These make for a good playground, but for a dangerous life. We avoid these things so much that if an organization were to take the form of a teeter-totter, we'd brace it up at both ends, turning it into a stable plank. But why has equilibrium become such a prized goal in adult life? Why do we seek so earnestly after balance? Is change so fearsome that we'll do anything to avoid it?

Sometimes, to clear up a confusing concept, it helps me to return to the accepted definition of the word. So I open the *American Heritage Dictionary* to learn about equilibrium: "Equilibrium. 1. A condition in which all acting influences are canceled by others resulting in a stable, balanced, or unchanging system. 2. Physics. The condition of a system in which the resultant of all acting forces is zero. . . . 4. Mental or emotional balance; poise."

I am surprised by the negativity of the first two definitions. A condition in which the result of all activity is zero? Why, then, do we desire equilibrium so much, or use the same word to describe mental and emotional well-being? In my own life, I don't experience equilibrium as an always desirable state. And I don't believe it is a desirable state for an organization. Quite the contrary. I've observed the search for organizational equilibrium as a sure path to institutional death, a road to zero trafficked by fearful people. Having noticed the negative effects of equilibrium so often, I've been puzzled why it has earned such high status. I now believe that it has to do with our outmoded views of thermodynamics.

Equilibrium is a result of the workings of the Second Law of Thermodynamics. Though we may not know what this law states, we act on its assumptions daily. My son learned it in fourth-grade physics as the "laziness law"—the tendency of closed systems to wear down, to give off energy that can never be retrieved. Ecologist Garrett Hardin aptly paraphrases this law: "We're sure to lose" (in Lovelock 1987, 124). Life goes on, but it's all downhill.

In classical thermodynamics, equilibrium is the end state in the evolution of closed systems, the point at which the system has exhausted all of its capacity for change, done its work, and dissipated its productive capacity into useless entropy. (Entropy is an inverse measure of a system's capacity for change. The more entropy there is, the less the system is capable of changing.) At equilibrium, there is nothing left for the system to do; it can produce nothing more. If the universe is a closed system (there being nothing outside the universe to influence it), then it too must eventually wind down and reach equilibrium. It will become a place where, in the words of scientists Peter

Coveney and Roger Highfield, "entropy and randomness are at their greatest, in which all life has died out" (1990, 153).

The Second Law of Thermodynamics applies only to isolated or closed systems—to machines, for example. The most obvious exception to this law is *life*. Everything alive is an open system that engages with its environment and continues to grow and evolve. Yet both our science and culture have been profoundly affected by the images of degeneration contained in classical thermodynamics. When we see decay as inevitable, or society as going to ruin, or time as the road to inexorable death, we are unintentional celebrants of the Second Law. James Lovelock, biologist and author of the Gaia hypothesis, says the laws of thermodynamics "read like the notice at the gates of Dante's Hell" (1987, 123).

If we believe that the universe is on a relentless road to death, we can't help but live in fear of change. In a downhill world, any change exhausts our store of valuable energy and leaves us empty, one step closer to death. Staying put or keeping our balance is a means of defense against the eroding forces of nature. We want nothing to change because only decline awaits us. Any form of present stasis is preferable to the known future of deterioration.

But in venerating equilibrium, we have blinded ourselves to the processes that foster life. It is both sad and ironic that we have treated organizations like machines, acting as though they were dead when all this time they've been living, open systems capable of self-renewal. We have magnified the tragedy by treating one another as machines, believing the only way we could motivate others was by pushing and prodding them into action, overcoming their inertia by the sheer force of our own energy. But here we are, living beings in living systems in a universe that continues to grow and evolve. Can we dump these thermodynamics and get to the heart of things? Can we respond to *life* in organizations and discard the death watch? Can we give up our clumsy attempts to keep things in balance and open ourselves to change?

Equilibrium is neither the goal nor the fate of living systems, simply because as open systems they partner with their environment. These systems are called "open" because they have the ability to continuously import energy from the environment and to export entropy. They don't sit quietly by as their energy dissipates. They don't seek equilibrium. Quite the opposite. To stay viable, open systems maintain a state of non-equilibrium, keeping themselves *off balance* so that the system can change and grow. They participate in an open exchange with their world, using what is there for their own growth. Every organism in nature, including us, behaves in this way.

In the past, systems analysts and scientists studied open systems primarily by focusing on the *structure* of the system (see Capra 1996, Part Two). This route led them away from observing or understanding the processes of change and growth that keep a system viable over time. Instead, analysts were interested in those influences that would support stability, which is the desired trait of machines. To maintain system stability, feedback loops were created to monitor what was going on. This type of feedback is called regulatory or negative feedback; it signals deviations from the established goal. Thermostats perform this function for heating systems. Managers perform a similar function when they evaluate performance against standard criteria, or compare progress against a plan. Negative or regulatory feedback helps keep a system on track once the course has been established. Information is used to help the system achieve its predetermined outcomes.

But there is a second type of feedback loop—positive or amplifying feedback. These loops use information differently, not to regulate, but to notice something new and amplify it into messages that signal a need to change. We recoil from the ear-piercing shrieks given off by microphones caught in a positive feedback loop. If stability, not growth, is the goal, then such amplification is very threatening, and we often rush in to quell it before eardrums burst. But positive feedback is essential to life's ability to adapt and change. In these loops, information increases and disturbances grow. The system, unable to deal with so much new and intensifying information, is being asked to change.

For many years, scientists failed to notice the role that positive feedback and disequilibrium played in facilitating a system's evolution. In trying to understand things as they were, in seeking to preserve system stability, they failed to note the internal processes by which open systems accomplish growth and change.

It was not until the element of time was introduced in Prigogine's study of thermodynamics that interest turned from system structures to *system dynamics*. His work, and those who developed it subsequently, dramatically expanded our awareness of how open systems use disequilibrium to avoid deterioration. Looking at the dynamics of open systems over time, scientists were able to see the effects of energy transformations they had not previously observed. Entropy, that fearful measure of a system's demise, was still being produced, sometimes in great quantities. But instead of simply measuring *how much* entropy was present, scientists could also note the dynamics of *what happened* to it—how quickly it was produced and whether it was exchanged with the environment.

Once it was noted that systems were capable of exchanging energy, trading usable energy for entropy, scientists realized that *deterioration was not inevitable*. Disturbances could create disequilibrium, but disequilibrium could lead to growth. If the system had the capacity to react and change, then disturbance was not necessarily a fearsome opponent. To understand the world from this perspective, scientists had to give up their views on decay and dissipation. They had to transform their ideas about the role of disequilibrium. They had to develop a new relationship with disorder.

Prigogine's work demonstrated that disequilibrium is the necessary condition for a system's growth. He named these systems *dissipative structures* to bring attention to their paradoxical nature. They dissipate or give up their form in order to recreate themselves into new forms. Faced with increasing levels of disturbance, these systems possess the innate ability to reorganize themselves to deal with the new information. For this reason, they are called *self-organizing systems*. They are adaptive and resilient rather than rigid and stable.

All life takes form as dissipative structures. Yet even in chemistry, with chemicals that are categorized as non-living, there are many startling examples of this self-organizing capacity. One example is a chemical clock, a solution that oscillates between two different states rather than existing as only one. In normal chemistry, when chemicals are mixed together, they form a solution in which the chemicals are evenly distributed. If a blue chemical is added to a red one, the resultant mixture will be purple. This is, in fact, the case in a chemical clock when it is at equilibrium and no reactions are taking place. But when change is introduced into this dissipative structure (in the form of new chemicals or changed conditions), the system is thrown into disequilibrium. It is then that the system behaves in a manner that defies normal expectations. Instead of purple, the substance begins to pulsate, first red, then blue, with predictability that earns them the title of "clock." To keep the clock-like pulsation going, the mixture must continue to be disturbed. If things stabilize and the disturbances cease, the pulsing stops and the solution settles into a static purple state. Equilibrium has returned, and there is nothing interesting left to see.

These chemical reactions use a great deal of energy. Entropy has increased during this reaction, but it has been exchanged for usable energy. As long as the system stays open to the environment and matter and energy continue to be exchanged, the system will avoid equilibrium and remain, instead, in these "evanescent structures" that exhibit "exquisitely ordered behavior" (Coveney and Highfield 1990, 164).

There are many examples of chemical reactions that exhibit extraordinary self-organizing behavior. One of the most beautiful is the Belousov–Zhabotinsky reaction, where chemicals, in response to changes in temperature and mix, form into swirling spiral patterns that rival the beauty of a Ukrainian Easter egg. The system is responding to disturbance by creating a new level of intricate organization. (See color section.)

The scrolls that emerge in the Belousov–Zhabotinsky reaction are similar to the scroll formations that appear in many places, both in nature and in art. "The spiral is one of nature's basic forms of design," writes photographer Andreas Feininger (1986, 124). Some scientists have wondered if spiral forms in art describe an archetypal experience of change, creation followed by dissipation and then new order. We see such spiral patterns in satellite photos of hurricanes. We live in a spiral-shaped galaxy; in fact, astronomers have concluded that the same iterative model used in the Belousov–Zhabotinsky chemical reaction applies to the scroll formation of star clusters. John Briggs, a science writer, and his writing colleague, physicist David Peat, describe the scroll images found so frequently in art, particularly noting the interlocking scroll patterns found in early motifs throughout the world: "Could such a collective wisdom perhaps be expressing its intuitions of the wholeness within nature, the order and simplicity, chance and predictability that lie in the interlocking and unfolding of things?" (1989, 142–43). (See color section.)

The self-organizing dynamics exhibited by these inert chemical solutions are evident in all open systems and in all life. These dynamics apply to such a broad spectrum of phenomena that they unify science across many disciplines. But, more importantly, they give us a new picture of the world; they "let us feel the *quality* of a world which gives birth to ever new variety and ever new manifestations of order against a background of constant change" (Jantsch 1980, 57).

I find the openness of self-organizing systems especially intriguing. Their relationship with their environment feels new to me. In organizations, we typically struggle against the environment, seeing it as the source of disruption and change. We tend to insulate ourselves from it as long as possible in an effort to preserve the precious stability we have acquired. Even though we know we need to be responsive to forces and demands beyond the boundaries of our organization, we still focus our efforts on maintaining the strongest defensive structure possible. We experience an inherent tension between stability and openness, a constant tug-of-war. But as I read about self-organizing systems, these dualities aren't present. Here are systems that stay strong by staying open. How do they do it?

The viability and resiliency of a self-organizing system comes from its great capacity to adapt as needed, to create structures that fit the moment. Neither form nor function alone dictates how the system is organized. Instead, they are *process structures,* reorganizing into different forms in order to maintain their identity. The system may maintain itself in its present form or evolve to a new order, depending on what is required. It is not locked into any one structure; it is capable of organizing into whatever form it determines best suits the present situation.

We are beginning to see organizations that are learning how to use the power of self-organization to be more agile and effective. There are increasing reports of organizations that have given up any reliance on permanent structures. They have eliminated rigidity—both physical and psychological—in order to support more fluid processes whereby temporary teams are created to deal with specific and ever-changing needs. They have simplified roles into minimal categories; they have knocked down walls and created workplaces where people, ideas, and information circulate freely. (See Petzinger, 1999)

At Oticon, a Scandinavian manufacturer of hearing aids, employees were given the freedom to redesign their physical space as part of a major destructuring of the entire corporate operation. They created maximum flexibility for themselves by foregoing offices or normal furniture. Employees created a nomadic office; each person received a cell phone, a laptop computer, and a file cart on wheels. As teams form, they wheel their file caddies up to neighboring tables and begin work. Their CEO tells the story of being gone from the office for a day, only to find his own rolling file cart wheeled into marketing. His staff had heard him mention that he needed to spend more time in that area (see Pinchot and Pinchot 1996).

If an organization seeks to develop these life-saving qualities of adaptability, it needs to open itself in many ways. Especially important is the organization's relationship to information, particularly to that which is new and even disturbing. Information must actively be sought from everywhere, from places and sources people never thought to look before. And then it must circulate freely so that many people can interpret it. The intent of this new information is to keep the system off-balance, alert to how it might need to change. An open organization doesn't look for information that makes it feel good, that verifies its past and validates its present. It is deliberately looking for information that might threaten its stability, knock it off balance, and open it to growth. This is so different from the way information is handled in well-defended organizations. In these, only information that confirms existing plans or leadership is let in. Closed off from disturbances, kept at equilibrium, such organizations run down, atrophy, and die (see also Chapter Six).

While a self-organizing system's openness to disequilibrium might seem to make it too unpredictable, even temperamental, this is not the case. Its stability comes from a deepening center, a clarity about who it is, what it needs, what is required to survive in its environment. Self-organizing systems are never passive, hapless victims, forced to react to their environments. As the system matures and develops self-knowledge, it becomes more adept at working with its environment. It uses available resources more effectively, sustaining and strengthening itself. It gradually develops a stability that then helps shelter it from many of the demands from the environment. This stability enables it to continue to develop in ways of its own choosing, not as a fearful reactant.

We see this pattern of development so clearly in ecosystems. In its early stages when the system is just forming, the early species that predominate produce large numbers of offspring. There is no system to provide stability or protection, so the production processes of these species are quite inefficient. They are very vulnerable to threats, and use a great deal of energy to produce many offspring, most of whom will be eaten. At this early stage, the environment exerts great pressure, playing a dominant role regarding which species survive. But as the ecosystem develops, created by relationships among many diverse species, a larger system emerges that is both stable and resilient. There is less pressure from the environment; therefore, species that use energy more efficiently can survive. Mammals, which produce far fewer offspring, can now flourish (see Jantsch 1980, 140ff; Margalef 1975). Even the environment changes, affected by its relationship with the ecosystem. Weather patterns, moisture levels, soil conditions—all are changed by the development of the ecosystem.

What occurs in these systems is contrary to our normal way of thinking. Openness to the environment over time spawns a stronger system, one that is less susceptible to externally induced change. What comes to dominate over time is not outside influences, but the self-organizing dynamics of the system itself. Because it partners *with* its environment, the system develops increasing autonomy *from* the environment and also develops new capacities that make it increasingly resourceful.

I say this is contrary thinking because we usually act from the reverse belief. We believe that in order to maintain ourselves and protect our individual freedom, we must defend ourselves from external forces. We tend to think that isolation, secrecy, and strong boundaries are the best way to preserve individuality. But this self-organizing world teaches that boundaries not only create distinctions; they are also places for communication and exchange (see Margulis and Sagan 1986). Because system members engage in continual exchanges among themselves and with their environment, the system develops greater freedom from its environment.

I have seen this paradox in action in a heavily regulated chemical manufacturing facility, the Dupont plant in Belle, West Virginia. As they opened their plant gates to government regulators, community people, schoolchildren, press, and even to environmental advocates, they gradually developed relationships with these diverse groups. Those relationships enabled them to engage together as learners and advocates. As trust developed and defensive postures faded, traditional boundaries dissolved. As plant manager Richard Knowles described this, "I no longer know where the plant ends, and I've learned it's not important to know that." As relationships developed far beyond the plant, it created conditions *within* the plant for levels of autonomy and experimentation that resulted in extraordinary new levels of safety and productivity.

A second process fundamental to all self-organizing systems is that of *self-reference*. When the environment shifts and the system notices that it needs to change, it always changes in such a way that it remains consistent with itself. This is autopoiesis in action, a system focused on maintaining itself, producing itself. It will choose a path into the future that it believes is congruent with who it has been. Change is never random; the system will not take off in bizarre new directions. Paradoxically, it is the system's need to maintain itself that may lead it to become something new and different. A living system changes in order to preserve itself.

Companies organized around a strong identity provide a good example of how self-reference works to create greater stability and autonomy (see Collins and Porras 1993; Blanchard and O'Connor 1997). When an organization knows who it is, what its strengths are, and what it is trying to accomplish, it can respond intelligently to changes from its environment. Whatever it decides to do is determined by this clear sense of self, not just because a new trend or market has appeared. The organization does not get locked into supporting certain products or business units just because they exist, or following after every fad just because it shows up. The presence of a clear identity makes the organization less vulnerable to its environment; it develops greater freedom to decide how it will respond.

Yet such companies are remarkably sensitive to their environment, staying wide open to new opportunities and ventures that welcome their particular skills. They also develop capacities to shape the environment, creating markets where none existed before. In the assessment of Prahalad and Hamel, companies focused on core competencies are able to "invent new markets, quickly enter emerging markets, and dramatically shift patterns of customer choice in established markets" (1990, 80; also, Hamel and Prahalad 1994).

Self-reference is the key to facilitating orderly change in the midst of turbulent environments. In organizations, just as with individuals, a clear sense of identity—the lens of values, traditions, history, dreams, experience, competencies, culture—is the only route to achieving independence from the environment. When the environment seems to demand a response, there is a means to interpret that demand. This prevents the vacillations, the constant reorganizations, and the frantic search for new customers and new ventures that continue to destroy so many businesses.

Another characteristic of self-organizing systems is their *stability over time*. Yet in describing them as stable, scientists are speaking about a quality of the global or whole system. Such global stability is maintained by another paradoxical situation, the presence of many local changes and instabilities occurring throughout the system. To use the example of an ecosystem again, any mature ecosystem experiences many changes and fluctuations among individuals and species. The total system achieves stability by supporting change within itself. Small, local disturbances are not suppressed; there is no central command function that stamps out these local fluctuations. It is by supporting them that the global system preserves its overall stability and integrity.

Jantsch notes the profound teaching embedded in these system characteristics: "The natural dynamics of simple dissipative structures teach the optimistic principle of which we tend to despair in the human world: *the more freedom in self-organization, the more order*" (1980, 40; italics added). This is, for me, the most illuminating paradox of all. The two forces that we have placed in opposition to one another—freedom and order—turn out to be partners in generating healthy, well-ordered systems. Effective self-organization is supported by two critical elements: a clear sense of identity, and freedom. In organizations, if people are free to make their own decisions, guided by a clear organizational identity for them to reference, the whole system develops greater coherence and strength. The organization is less controlling, but more orderly.

In addition to these tantalizing paradoxes, self-organizing systems teach an important lesson about how change happens in living systems. When the system is far from equilibrium, singular or small influences can have enormous impact. It is not the law of large numbers or critical mass that creates change, but the presence of a small disturbance that gets into the system and is then amplified through the networks. Once inside the network, this small disturbance circulates and feeds back on itself. As different parts of the system get hold of it, interpret it, and change it, the disturbance grows. Finally, it becomes so amplified that it cannot be ignored. We've all had this experience, probably more than once: A casual or offhanded comment tossed out in a meeting gets picked up by the organization, and suddenly we're in the midst of a firestorm of opinions, emotions, and rumors.

Whenever a self-organizing system experiences any amplification process, change is at hand. If the amplifications increase to the level where they destabilize the system, the system can no longer remain as it is. At this moment, the system is at a crossroads, standing poised between death and transformation. In science, this is known technically as a bifurcation point. For us humans, it is known as a moment of great fear, tinged, perhaps, with a faint sense of expectation. At this point, the system's future is wide open. Abandoning its present form, the system is free to seek out a new form in response to the changed environment. Even the forces of evolution are not constraining. Self-reference will be at work, but otherwise the system has no predetermined course. At the bifurcation point, "such systems seem to 'hesitate' among various possible directions of evolution," Prigogine and Stengers state; "a small fluctuation may start an entirely new evolution that will drastically change the whole behavior of the macroscopic system" (1984, 14).

I can think of several organizations, particularly customer-oriented ones, that brag about how a single customer inquiry or the suggestion of one employee directed them into entirely new product lines that became very successful. There was no preplanning, no long-range strategic objectives, that led them into these markets. Just the creativity of one or a few individuals who succeeded in getting the attention of the organization and then watched the suggestion amplify to the level where the company reorganized to respond to it.

In describing self-organization, I am always struck by the great partnering that exists between the system and its environment. As the system changes and evolves, it also affects its environment. No participant in this dance is left unaffected by changes that occur in another. Scientists call this *co-evolution*. Organizational theorist William Starbuck wrote about this process in organizations years ago. The constraints imposed by the environment, he noted, do not force the organization to behave a certain way: "Organizations and their environments are evolving simultaneously toward better fitness for each other" (1976, 1105–6). In this view of evolution, the system changes, the environment changes, and even the rules of evolution change: "Evolution is the result of self-transcendence at all levels. . . . [It] is basically open. It determines its own dynamics and direction. . . . By way of this dynamic interconnectedness, evolution also determines its own *meaning*" (Jantsch 1980, 14).

All life lives off-balance in a world that is open to change. And all of life is self-organizing. We do not have to fear disequilibrium, nor do we have to approach change so fearfully. Instead, we can realize that, like all life, we know how to grow and evolve in the midst of constant flux. There is a path through change that leads to greater independence and resiliency. We dance along this path by maintaining a coherent identity and by honoring everybody's need for self-determination.

When leaders strive for equilibrium and stability by imposing control, constricting people's freedom and inhibiting local change, they only create the conditions that threaten the organization's survival. We have all experienced this in our organizational lives, but to understand how dangerous it is to restrict fluctuations and change, we can look again at human experience with ecosystems. How many ecological messes have we had to cope with because of management practices that sought to preserve wilderness by protecting it from small, natural fluctuations, or by eliminating predators? In Yellowstone National Park, human-imposed stability thwarted for many years the natural process of small fires, which regularly clean out brush and dead trees. The result was a fragile equilibrium completely vulnerable to the cataclysm of fire that destroyed large areas of the park. The attempt to manage for stability and to enforce an unnatural equilibrium always leads to far-reaching destruction.

The more I read about self-organizing systems, the more I marvel at the images of freedom and possibility they evoke. This is a world of independence and interdependence, of processes that resolve so many of the dualisms we created in thought. The seeming paradoxes of order and freedom, of being and becoming, whirl into a new image that is very ancient—the unifying spiral dance of creation. Stasis, balance, equilibrium, these are temporary states. What endures is process—dynamic, adaptive, creative.

Self-organizing systems offer compelling lessons in how the world works, of how order is sustained in the midst of change. This is very new territory for us, and it is hard to silence our well-trained linear minds, to stop grasping for small ideas and techniques that we can apply immediately to our work. But before we reach for applications, I hope we are willing to sit quietly and contemplate the great paradoxes of this new land. Let us not move too quickly across its features, heads down, blinded by our past beliefs, looking only for some small ways to use this knowledge right now. Instead, let us stand still for a moment and dwell in the realization that we live in a world of inherent order, where paradoxical but natural processes exist for growth and self-renewal.

I find pleasure in letting these new ideas swirl freely inside me. Like clouds, they begin as mist, then take form, then dissipate. Clouds themselves are self-organizing, taking new shape as thunderstorms, hurricanes, or rain fronts depending on changes in their environment. We are capable of similar transformations; new ideas can emerge as powerful insights if we allow them the freedom to self-organize. And there is much we can learn from clouds. They are spectacular examples of fluid and responsive systems, structured in ways we never imagined possible: "After all, how do you hold a hundred tons of water in the air with no visible means of support? You build a cloud" (Cole 1985, 38).

Whether an order is formed or not depends on whether or not information is created . . . the essence of creating order is in the creation of information.

—*Ikujiro Nonaka*

Chapter 6

The Creative Energy of the Universe —Information

Why is there such an epidemic of "poor communications" within organizations? In every one I've been in, employees have ranked it right at the top of their major issues. Indeed, its appearance on those lists became so predictable that I grew somewhat numb to it. Poor communication was a superficial diagnosis, I thought, that covered up other, more specific issues. Over the years, I developed a conditioned response to "communications problems" the minute they were brought up. I disregarded the assessment. I started pushing people to "get beyond" that catch-all phrase, to "give me more concrete examples" of communications failures. I believed I was en route to the "real" issues that would have nothing to do with communication.

Now I know I was wrong. My frustration with pat phrases didn't arise from people's lack of clarity about what was bothering them. They were right. They were suffering from problems related to information. Asking them to identify smaller, more specific issues was pushing them in exactly the wrong direction, because the real problems were big—bigger than anything I imagined. What we were all suffering from, then and now, is a fundamental misperception of information: what it is, how it behaves, how to work with it.

The nub of the problem is that we've treated information as a "thing," as a physical entity. A "thing" has material form; you can get your hands around it, move it from place to place, expect to pass it on unchanged. You can manage things.

For several decades, information theory has treated information as something this tangible. Information has been referred to as a quantity, bits and bytes to be counted, transmitted, received, and stored. Information is a commodity that we transfer from one place to another. We maintain this commodity focus even now when we evaluate the conductivity of a transmission line, or a computer's capacity, by calculating how much information it can hold. This strong focus on the "thingness" of information has kept us from contemplating its other dimensions: the content, character, and behavior of information (Gleick 1987, 255–56). Information technology still has as a primary concern the smooth, uninterrupted transmission of information. Engineers and leaders alike still hope that information can move virgin-like through the system, untouched by anything.

I believe it is this history with information theory that has gotten us into trouble. We don't understand information at all.

What's curious about our misperception of information is that we all started out on a much higher plane of awareness. Remember playing "telephone" and being delighted and amazed at how the message got distorted as it was whispered from ear to ear? At a young age, we were charmed by information's dynamic nature, by its unpredictable and constantly changing character. But when we entered organizational life, we forgot that experience. We expected information to be controllable, stable, and obedient. We expected to be able to manage it.

In the universe that new science is exploring, information is a very different "thing." It is not the limited, quantifiable, put-it-in-an-e-mail-and-send commodity that we pretend it to be. In new theories of evolution and order, information is a dynamic, changing element, taking center stage. Without information, life cannot give birth to anything new; information is absolutely essential for the emergence of new order.

All life uses information to organize itself into form. A living being is not a stable structure, but a continuous *process* of organizing information. A dramatic example of this, one that pushes our self-concept to the edge, is demonstrated by asking: Who am I? Am I a physical structure that processes information or immaterial information organizing itself into material form?

Although we experience ourselves as stable form, our body changes frequently. As physician/philosopher Deepak Chopra likes to explain, our skin is new every month, our liver every six weeks; even our brain, with all those valuable cells, changes its content of carbon, nitrogen, and oxygen about every twelve months. Day after day, as we inhale and exhale, we give off what were our cells and take in elements from other organisms to create new cells. "All of us," observes Chopra, "are much more like a river than anything frozen in time and space" (1990). In spite of this exchange of physical matter, we remain rather constant, due to the organizing function of the information contained in our bodies:

> At any point in the bodymind, two things come together—a bit of information and a bit of matter. Of the two, *the information has a longer life span than the solid matter it is matched with. . . .* This fact makes us realize that memory must be more permanent than matter. What is a cell, then? *It is a memory that has built some matter around itself, forming a specific pattern. Your body is just the place your memory calls home.* (Chopra 1989, 87; italics added)

Jantsch describes the same phenomenon in all life, asking whether a self-organizing system is best understood as a material structure that organizes energy or as information processes that organize the flow of matter. He concludes that self-organizing systems are better thought of as energy processes that manifest themselves as physical forms (1980, 35). And biologist Steven Rose develops important questions from the same conclusion: "Organisms have forms which change but also persist throughout their life's trajectory, despite the fact that every molecule in their body has been replaced thousands of times over during their lifetime. How is form achieved and maintained? What are living organisms made of?" (1997, 16).

Life uses information to organize matter into form, resulting in all the physical structures that we see. The role of information is revealed in the word itself: *in-formation*. We haven't noticed information as integral to the process of formation because all around us are physical forms that we can see and touch. These things beguile us; we confuse the system's physical manifestation with the processes that gave birth to it. Yet the real system, that which endures and evolves, is a set of processes. Information takes shape in different forms as a result of these processes. When a new structure materializes, we know that the system has in-formed itself differently.

In a constantly evolving, dynamic universe, information is a fundamental yet invisible player, one we can't see until it takes physical form. Something we cannot see, touch, or get our hands on is out there, influencing life. Information seems to be managing us.

For a system to remain alive, for the universe to keep growing, information must be continually generated. If there is nothing new, or if the information merely confirms what already is, then the result will be death. Closed systems wind down and decay, victims of the Second Law of Thermodynamics. The source of life is new information—novelty—ordered into new structures. We need to have information coursing through our systems, disturbing the peace, imbuing everything it touches with the possibility of new life. We need, therefore, to develop new approaches to information—not management but encouragement, not control but genesis. How do we create more of this wonderful life source?

Information is unique as a resource because it can generate itself. It's the solar energy of organization—inexhaustible, with new progeny possible with every interpretation. As long as communication occurs in a shared context, fertility abounds. These new births require freedom; information must be free to circulate and find new partners. The greatest generator of information is the freedom of chaos, where every moment is new. With so much spawning going on, scientists feel obliged to watch carefully a chaotic system's activity lest they miss something (Gleick 1987, 260).

Of course, such freedom is exactly what we try to prevent. We have no desire to let information roam about promiscuously, procreating where it will, creating chaos. Management's task is to enforce control, to keep information contained, to pass it down in such a way that no newness occurs. Information chastity belts are a central management function. The last thing we need is information running loose in our organizations. And there are good reasons for our stern, puritanical attitudes toward information; unfettered information has created enough horror stories to justify frequent witch hunts.

But if information is to function as a source of organizational vitality, we must abandon our dark cloaks of control and trust in its need for free movement, even in our own organizations. Information is necessary for new order, an order we do not impose, but order nonetheless. All of life uses information this way. Can information, then, be used as a helpmate in creating greater order in our organizations?

Information can serve such an organizational function because organizations are open systems and are responsive to the same self-organizing dynamics as all other life. To foster these self-organizing capacities in our organizations, we have to work with information the same way that life does.

We have to create much freer access to it, and become much more astute at noticing new information as it emerges. No other species seems to suffer from the delusion that they can manage information. Instead, they stay alert to what's happening all the time. It seems ironic that even the simplest forms of life often seem more self-aware than we humans do. In many fields of science, we glimpse how life uses the information it gathers not just to preserve itself, but to grow and generate new capacities.

Prigogine was stimulated to think about such issues when he observed a process of communication even in "non-living" chemical reactions. He came to the rather startling conclusion that in certain inanimate chemical solutions, the molecules were communicating with one another to generate new order. In the chemical clocks he studied, at a certain point the random mix of molecules becomes coordinated. A murky dull solution, for example, suddenly begins pulsing, first blue, then clear. The molecules act in total synchronization, changing their chemical identity simultaneously. "The amazing thing," Prigogine notes, "is that each molecule knows in some way what the other molecules will do at the same time, over relatively macroscopic distances. These experiments provide examples of the ways in which molecules communicate. . . . That is a property everybody always accepted in living systems, but in nonliving systems it was quite unexpected" (1983, 90).

If a system has the capacity to process information, to notice and respond, then that system possesses the quality of *intelligence*. It has the means to recognize and interpret what is going on around it. Researchers working in artificial life suggest that intelligence can't be discerned from noting the constituent parts of an entity (see Kelly 1994). An organism doesn't even need a brain in order to be intelligent. Intelligence is a property that emerges when a certain level of organization is reached which enables the system to process information. The greater the ability to process information, the greater the level of intelligence.

Gregory Bateson (1980) specified similar criteria in defining "mind." Any entity that has capacities for generating and absorbing information, for feedback, for self-regulation, possesses mind. These definitions offer us a means to contemplate organizational intelligence: why some organizations seem so smart, why others fail to survive for long, and why still others get stuck in repeating the same mistakes. We can begin to see that organizational intelligence is not something that resides in a few experts, specialists, or leaders. Instead, it is a system-wide capacity directly related to how open the organization is to new and disconfirming information, and how effectively that information can be interpreted by anyone in the organization.

Everybody needs information to do their work. We are so needy of this resource that if we can't get the real thing, we make it up. When rumors proliferate and gossip gets out of hand, it is always a sign that people lack the genuine article—honest, meaningful information. Given that we all need to be continually nourished by information, it is no wonder that employees cite "poor communication" as one of their greatest problems. People know it is critical to their ability to do good work. They know when they are starving.

We have lived for so long in the tight confines of bureaucracies—what Max De Pree, former CEO of Herman Miller, describes as "the most superficial and fatuous of all relationships"—that it is taking us some time to learn how to live in open, intelligent organizations. This requires an entirely new relationship with information, one in which we embrace its living properties. Not so that we open ourselves to indiscriminate chaos, but so that we can facilitate effective responses in a world that is constantly surprising us. If we are seeking *resilient* organizations, a prized property of living systems, information is a key ally.

Think about how we generally have treated information. We've known it was important, but we've handled it in ways that have destroyed many of its life-giving properties. For one thing, we haven't been interested in newness. We've taken disturbances and fluctuations and averaged them together to give us comfortable statistics. Our training has been to look for large numbers, important trends, major variances. We live in a society that believes it can define *normal* and then judge everything against this fictitious standard. We struggle to smooth out the differences, conform to standards, measure up. Yet in life, newness can only show up as difference. If we aren't looking for differences, we can't see that anything has changed; consequently, we aren't able to respond.

Even when we do notice new information, we too often rush in to kill it off. Instead of appreciating the rich possibilities that could move us to new levels of understanding, we think we're wise enough to play instant Solomon. We don't want to dwell in confusion. We value quick decisions over wise ones. "Let's get this over with," we say. "Let's just make a decision." We aim our efforts dead into solid ground, away from the exploration that would move us toward the light of richer understanding. For so long, we've been engaged in smoothing things over, rounding things off, keeping the lid on (the metaphors are numerous), that our organizations have literally been dying for information they could feed on, information that was different, disconfirming, and filled with enough newness to disturb the system into wise solutions.

We do not exist at the whim of information; that is not the fearsome prospect which greets us in a world ravenous for information. Our own capacity for meaning-making plays a crucial role. We, alone and in groups, serve as interpreters, deciding which information to pay attention to, which to suppress. We are already highly skilled at this, but we would benefit from noticing just how much interpretation we do, and how we might develop new lenses of discernment. We can open ourselves to more information, in more places, and seek out that which is ambiguous, complex, perhaps even irrelevant. I know one organization that thinks of information as salmon. If its organizational streams are well-stocked, they believe, information will find its way to where it needs to be. It will swim upstream to where it can spawn. The organization's job is to keep the streams clear so that information has an easier time of it. The result is a harvest of new ideas and projects.

Another organization was able to change its approach to information by changing its metaphors. Instead of the limiting thought that "information is power," they began to think of information as "nourishment." This shift keeps their attention on the fact that information is essential to everyone, and that those who have more of it will be more intelligent workers than those who are starving.

Information is always spawned out of uncertain, even chaotic circumstances. This is not a reassuring prospect. How are we to welcome information into our organizations and ally ourselves with it as a partner in our search for order if the processes that give it birth are ambiguity and surprise? In a profession that has raised the practice of "no surprises" to a high art, sponsoring such processes reads like a macabre prescription for self-destruction. Few things make us more frantic than increasing ambiguity. And although we say we've come to tolerate ambiguity rather well over the past years (because we had no other choice—it wasn't going away), it often appears that we don't tolerate it as much as we shield ourselves from it. We have a hard time with lack of clarity, or with questions that have no easy answers. We move hurriedly out of these discomforts by focusing on one element, coming up with a narrow solution, and pretending not to notice everything else that's not getting dealt with. We feel safer with blinders on, fearing that if we open our eyes this will only add to our distress, even though our experience suggests that when we keep ourselves blinded, we are more frequently "blindsided."

We refuse to accept ambiguity and surprise as part of life because we hold onto the myth that prediction and control are possible. We still believe that it is possible to control every part of the machine. We still believe that we can (and must) know what's going on everywhere. We still believe that what holds a system together is us, our leadership. It is *our* intelligence—not the intelligence distributed broadly throughout the organization—that brings order to everything. When things start to feel confusing or ambiguous, no wonder we get anxious. Ambiguity asks us to contemplate even more variables, confusion asks us to say that we don't know. We know we can't possibly gain control of even more elements, stressed and stretched as we already are. Our span of control pulls away from us elastically, and, suddenly, we are catapulted into unmanageability. Under such pressure, it's no wonder that we want to shut out newness and hold on blindly to the few things that worked in the past.

But there is a way out of the paralyzing fear that ambiguity engenders. It requires that we step back, refocus our attention on the system as a whole, and realize there are other processes at work. Beyond our leadership skills, and often in spite of them, the system is self-organizing to accomplish its work.

This is such a remarkably different perspective, and it calls for new skills in us. We all have to learn how to support the workings of each other, to realize that intelligence is distributed and that it is our role to nourish others with truthful, meaningful information. Fed by such information, everyone can more capably deal with issues and dilemmas that appear in their area. It is no longer the leader's task to deal with all problems piece by piece, in a linear and never satisfying fashion. It is no longer the leader's task to move information carefully along restricted pathways, shepherding it cautiously through channels, passing it on guardedly to someone else. This was how leaders were taught to manage in the past. And mechanistic models of brain function reinforced this as the correct approach. Earlier brain physiology described information as moving step by step from one neuron to the next, just as methodically as leaders have tried to do. But brain function is now described with imagery that bears no resemblance to these mechanistic notions of the past. These new ideas offer many possibilities for more open and liberated ways of distributing information.

In newer theories of the brain, information is widely distributed, not necessarily limited to specific neuron sites. In mapping areas of the brain to determine those that relate to specific signals (for example, those related to hand movements), neuroscientists have found that these "sites" do not correspond to any particular neurons. Instead of a specific physical place, they observe a more fluid pattern of electrical activity. Instructions, such as those for a particular finger movement, seem to be distributed through a shifting network. And memories, it is now thought, "must arise as relationships within the whole neural network" (Briggs and Peat 1989, 171). If information is stored in these *networks of relationships* among neurons, damage to a particular area of the brain will not result in the loss of that information. Other areas in the network may retain that information in some form.

These *neural nets* were first simulated in meager degree by assembling more than 60,000 computers and linking them together to do parallel processing. Zohar describes the brain's neural net as a "rather messy, higgledy-piggledy wiring design, where everything seems randomly connected to everything else" (1990, 72). In our brains—and the computers that can never hope to mimic them—complex information travels across broad expanses, never organized into neat pathways, yet capable of organizing into memory and functions.

Instead of channeled flows of information, neural nets give us images of information moving in all directions simultaneously. How this rather "higgledy-piggledy" system works is not clear. Scientists can neither precisely track nor control how such random distribution of information achieves a sense-making capacity. But we each live inside bodies where we rely on the effectiveness of these processes.

Several years ago, a major long-distance phone company discovered that telephone calls could be routed more efficiently and effectively anywhere on the globe if the routing was not controlled by a centralized unit. In place of centralized decisions, they created the technology to support a rapid exchange of information among the various switches. Each call could find its own best route by quickly scanning what was going on in the system. However, as one manager sadly reported, at the same time that his company was discovering how well this worked for machines, it had yet to trust that similar processes would lead to far improved functioning at the human level among employees.

We have many organizational models that demonstrate how open access to information contributes to self-organized effectiveness. The literature on organizational innovation, creativity, and knowledge management is rich in lessons that apply here; not surprisingly, they describe processes that also characterize the natural universe. Innovation is fostered by information gathered from new connections; from insights gained by journeys into other disciplines or places; from active, collegial networks and fluid, open boundaries. Knowledge grows inside relationships, from ongoing circles of exchange where information is not just accumulated by individuals, but is willingly shared. Information-rich, ambiguous environments are the source of surprising new births.

We need look no farther than our computer screens to see how open information contributes to our personal effectiveness and knowledge. The Internet gives us full access to information formerly held by a few. In the early years of the Internet, medical doctors reported that their patients, who researched their conditions on-line, knew more about their treatment options than they did. This was a disturbing shift for physicians—partnering with their patients rather being in charge. Now, many healthcare systems expect patients to add their own research to that done by physicians. And think about how much more effective you feel in negotiating loans or major merchandise purchases because you know from the Web what's going on in the market. Our lives are dramatically different because we can search the Web and find the information we need instantly (see Locke, et.al. 2001, Weinberger 2002).

A very different process for how new and abundant information can facilitate self-organization is found in organizational change work described as

"Whole Systems" (see Holman and Devane 1999). One model now in wide use, is "Future Search" (see Weisbord and Janoff 1995). The whole system—sometimes literally, sometimes through selected members—is gathered in one room to develop a desired future for the organization. People from all parts of the organization, including those "outsiders" who in truth are very connected to it, work together to generate information on the organization's history, its present capacities, and its external demands. The first day is spent bringing to the surface the information contained in the organizational neural net—opinions, interpretations, and history carried within all the different people in the room. Information is generated in deliberately overwhelming amounts.

In the presence of so much information, people often feel temporarily powerless and disheartened. They don't know how to make sense of it, and they are in that terribly uncomfortable state of feeling confused. But as information continues to proliferate and confusion grows, there comes a memorable time (usually during the last quarter of the event) when the group self-organizes, growing all that information into new, potent visions of the future. Rather than basing agreements on the lowest common denominator, the whole system that is present at the conference has self-organized into a new creation, a unified body that sets new and challenging directions for itself.

Although overwhelming levels of information are intentionally created in these sessions, it is never the volume that matters. It is only the *meaning* of information that makes it potent or not. When information is identified as meaningful, it is a force for change. In the system's networks and feedback loops, such information circulates and grows and mutates in the conversations and interactions that occur. This process seems to be the way nature creates the well-ordered and diverse beauty that delights us: Information is generated freely by the system and fed back on itself so that it continues to grow and change.

It is just such a process that gives birth to the ineffable beauty of fractals (see color section). These geometrical forms are generated by computers from relatively little information that is expressed in a few nonlinear equations. The equations are not there to be solved just once; instead, each solution is a contribution to the creation of a complex pattern. As one solution is found, it is immediately fed back into the equation so that another, different solution can develop. This process has been termed "evolving feedback." As the equations are fed back on themselves, evolving a new solution with every iteration, elaborate levels of pattern and differentiation are created. These patterns never end; as long as the iterative process continues, the patterns will continue to evolve into infinity:

> Fractals are . . . complex by virtue of their infinite detail and unique mathematical properties (no two fractals are the same), yet they're simple because they can be generated through successive applications of simple iterations. . . . It's a new brand of reductionism . . . utterly unlike the old reductionism, which sees complexity as built up out of simple forms, as an intricate building is made out of a few simple shapes or bricks. *Here the simple iteration in effect liberates the complexity hidden within it, giving access to creative potential.* The equation isn't the plot of a shape as it is in Euclid. Rather, the equation provides the starting point for evolving feedback. (Briggs and Peat 1989, 104; italics added)

The process of fractal creation suggests some ways organizations can work with the paradox that greater openness is the path to greater order. A fractal reveals its complex shape through continuous self-reference to a simple initial equation. Thus, the work of any team or organization needs to start with a clear sense of what they are trying to accomplish and how they want to behave together. I think of these agreements as the initial equations (see also Chapter Seven). Once this clarity is established, people will use it as their lens to interpret information, surprises, experience. They will be able to figure out what and how to do their work. Their individual decisions will not look the same, and there is no need for conformity in their behavior. But over time, as their individual solutions are fed back into the system, as learning is shared, we can expect that an orderly pattern will emerge.

At all levels and for all activities in organizations, we need to challenge ourselves to create greater access to information and to reduce those control functions that restrict its flow. We cannot continue to use information technology and management systems as gatekeepers, excluding and predefining who needs to know what. Instead, we need to evoke contribution through freedom, trusting that people can make sense of the information because they know their jobs, and they know the organizational or team purpose. Restricting information and carefully guarding it doesn't make us good managers. It just stops good people from doing good work. Jan Carlson, former head of Scandinavian Airlines and one of the pioneers in the customer service revolution, says it clearly: "An individual without information cannot take responsibility, but an individual who is given information cannot help but take responsibility" (Willett 1999). Information provides true nourishment; it enables people to do their jobs responsibly and well.

Probably the most startling example of an organization that is redesigning itself because of increased access to information is the U.S. armed forces. Both the Army and the Marines now have the technology to provide every individual soldier with information about what's occurring on the battlefield, information that formerly was known only by the commanders. Through extensive field tests, the Army has discovered that when individuals have such information and know how to interpret it because they know the "commander's intent," they can make decisions that lead to greater success in battle. They respond quickly and intelligently, and assume responsibility for their decisions. Although it has been difficult for some older commanders to turn over so much control, the evidence is very clear that a network form of organization, where people are linked together by technology and shared meaning, makes soldiers more effective. Because of this demonstrated effectiveness, the Army and Marines have announced that they are moving into a networked form of command quite different from their historical traditions.

As this Army story illustrates, an organization that wants to learn has to be willing to look at information that disconfirms its past beliefs and practices. Organizations that want to stay vital must search out surprise, looking for what is startling, uncomfortable, and maybe even shocking. The organization then needs to support people to reflect on this unsettling or disconfirming information, providing them with the resources of time, colleagues, and reflection. The value of this has been evident in processes such as scenario planning, and some approaches to quality and knowledge management. People are encouraged to look for variances, to travel far afield and bring in newness. They are encouraged to think together to decide what the information means.

Anything that supports reflective conversations among new and different parts of the organization is important, including architectural spaces for informal exchanges and dedicated time in meetings. Through these processes, new information is spawned, new meanings develop, and the organization grows in intelligence. I am intrigued by the thought that these programs work well not simply because they invite employee contribution and involvement, but because they generate more of the very substance that is required to reorder the universe—new information.

Jantsch, as a scientist, urges managers to a new role, that of "equilibrium busters." No longer the caretakers of control, we become the grand disturbers. We stir things up and roil the pot, looking always to provoke, even to disrupt, until finally things become so confusing that the system must reorganize itself into new forms and new behaviors. If we accept this challenge to be equilibrium busters, if we begin to value that it is disequilibrium that keeps us alive, we will find the task quite easy. There is more than enough confusion and ambiguity in our lives to work with. We don't have to worry about creating more, only about how to work more artfully with what we already have.

Who doesn't feel confused these days, or overwhelmed and overloaded by so much information? I think it's important to remember that we are only infants in learning how to deal with the volume of information that technology makes available. The analytic thought processes we learned in school and business have not prepared us to deal with the quantity of information that bombards us. Many creativity teachers suggest that we use such a small part of our mental capacity because of our insistence on linear thinking. We can't use neat and incremental methods to make sense of the world any longer. We need to be experimenting with thinking processes that better suit our neural netlike brains, those processes that are open, nonlinear, messy, relational. As we develop these, we will learn new ways to process the mass of information that too often overwhelms us now. As we learn to deal with information on its own terms, we will come to treasure it as the essential partner that it is.

It is not only individuals who have to become more creative and think "outside the box." Organizations too must move beyond the boxes they have drawn to describe roles and relationships. Many organizations are experimenting with new organization charts that describe more fluid patterns of relationship. While none of these quite succeed in describing the true complexity of the relationships, each attempts to communicate a more accurate picture of organizational life. Francis Hesselbein, Chairman of the Leader to Leader Institute, believes we are again learning "to manage in a world that is round," a world not of hierarchies but of encircling partnerships (Hesselbein and Cohen 1999, Ch. 2). Buckman Labs is moving from "a chain of command to a web of influence" (Willett 1999, 2). And Gore Associates, manufacturers of GoreTex®, describes itself as a "lattice organization." These images describe organizations where roles and structure are created from need and interest, where relationships among workers are nurtured as the primary source of organizational creativity and success. One observer of Gore has noted that the issue is not who or what position will take care of the problem, but what energy, skill, influence, and wisdom are available to contribute to the solution (Pacanowski 1988).

Many organizations are struggling with how to use information to become more intelligent. Thinking has been acknowledged as a critical skill, and not just at higher levels of management. It is now recognized that many more workers need to be able to interpret complex information. Information and thinking skills that formerly were the purview of the leader are moving deeper into organizations. This work comes under different banners: Learning Organization, Business Literacy, Intellectual Capital, Knowledge Management. Each assumes that intelligence must be broadly distributed. As Gifford Pinchot states, "The measure of organizational intelligence is quite straightforward. It's one brain per person." When Buckman Labs set out to develop greater organizational knowledge, they set as their challenge how to create access to the information that was distributed across more than twelve hundred minds working in twenty-one different countries (Willett 1999).

One of an organization's most critical competencies is to create the conditions that both generate new knowledge and help it to be freely shared. More and more, there is an acknowledged benefit to sharing information within and beyond the organization, to doing away with the gates and blockages, to moving past the hoarding and the fear, to developing trusting relationships. Does this mean we can expect greater organizational intelligence?

My own faith that organizations are evolving to greater intelligence comes from my understanding that we live in an intrinsically well-ordered universe. As I read further into new science, I recognize that living systems engage with life differently than we do. We struggle to carefully build order, layer upon layer, while life's order emerges. We labor hard to hold things together, while life participates together openly and self-organized structures emerge. Jantsch contrasts our traditional approach of building block by block to nature's process of "unfolding" (1980, 75). From the "interweaving of processes" new capacities and structures emerge. Order is never imposed from the top down or from the outside in. Order emerges as elements of the system work together, discovering each other and together inventing new capacities.

We need to learn more about these sources of order. In ways we have failed to notice, systems possess the capacity to self-organize. As we learn to work with this ability, our attention will shift away from the *parts*, those rusting holdovers from an earlier age of organizing, and focus us on the deeper, embedded processes that create effective organizations. "What is needed," writes Bohm, "is an *act of understanding* in which we see the totality as an actual process that, when carried out properly, tends to bring about a harmonious and orderly overall action, in which analysis into parts has no meaning" (1980, 56).

In quantum physics, a homologous process is described as *relational holism,* where whole systems are created by the relationships among subatomic particles. In this process, the parts don't remain as parts; they are drawn together by a process of internal connectedness. Electrons are drawn into these intimate relations as they cross paths with one another, overlapping and merging; their own individual qualities become indistinguishable: "The whole will, as a whole, possess a definite mass, charge, spin, and so on, but it is completely indeterminate which electrons are contributing what to this. Indeed, it is no longer meaningful to talk of the constituent electrons' individual properties, as these continually change to meet the requirements of the whole" (Zohar 1990, 99).

This is an intriguing image for organizations. It is not difficult to recognize ourselves as electrons in organizations, moving, merging with others, forming new wholes, being forever changed in the process. We experience this when we say that a team has "jelled," suddenly able to work in harmony, the ragged edges gone, an effortless flow to the work. We all have experienced things "coming together," or been in team efforts that far exceeded what we could do alone, but these have always felt slightly miraculous. We never understood that we were participants in a universe that thrives on open information and that works with us to self-organize into systems of increased capacity.

We speak more these days about fluid and permeable boundaries; we know that organizations have to be more open to meet the unending pressures for change. The notion of permeable boundaries has sparked both fear and curiosity. Perhaps if we understand the deep support we have from natural processes, it will help dispel some of the fear. It is not that we are moving toward disorder when we dissolve current structures and speak of worlds without boundaries. Rather, we are engaging in a fundamentally new relationship with order, order that is identified in processes that manifest themselves only temporarily as structures. Order itself is not rigid or located in any one structure; it is a dynamic organizing energy. When this organizing energy is nourished by information, we are given the gifts of the living universe. The gift is evolution, growth into new forms. Life goes on, richer, more creative than before.

Thus before all else, there came into being the Gaping Chasm, Chaos, but there followed the broad-chested Earth, Gaia, the forever-secure seat of the immortals . . . and also Love, Eros, the most beautiful of the immortal gods, he who breaks limbs

—Hesiod

Chapter 7

Chaos and the Strange Attractor of Meaning

S everal thousand years ago, when primal forces haunted human imagination, great gods arose in myths to explain the creation of the world. At the beginning was Chaos, the endless, yawning chasm devoid of form or fullness. And there also was Gaia, mother of the earth, she who brought forth form and stability. In Greek story, Chaos and Gaia were partners, two primordial powers engaged in a duet of opposition and resonance, creating everything we know.

These two mythic figures again inhabit our imagination and our science. They have taken on new life as scientists explore more deeply the workings of our universe. For me, this return to mythic wisdom is both intriguing and comforting. It signifies that even as we live in the midst of increasing turbulence, a new relationship with Chaos is possible. Like ancient Gaia, we are being asked to partner with Chaos, understanding it as the life process that releases our creative power. From Chaos' great chasm comes both support and opposition, creating the "light without which no form would be visible" (Bonnefoy 1991, 369–70). We, the generative force, give birth to form and meaning, organizing Chaos through our creativity. We fill the void with worlds of our own making and turn our backs on him. But we must remember that deep within our Gaian centers, so the Greeks and our science tell us, is the necessary heart of Chaos.

The heart of chaos has been revealed with modern computers. Watching the behavior of a chaotic system as it is tracked on a computer screen is a mesmerizing experience. The computer records the evolution of the system, displaying each moment of the system's chaotic behavior as a point of light on the screen. Because of the computer's speed, we can soon observe how the system is evolving. The system careens back and forth with raucous unpredictability, never showing up in the same spot twice. But as we watch, this chaotic behavior weaves into a pattern, and before our eyes order emerges on the screen. The chaotic movements of the system have formed themselves into a shape. The shape is a "strange attractor," and what has appeared on the screen is the order inherent in chaos (see illustration on the next page).

Strange attractors evoke feelings of awe in most who observe them. Poetic language frequently creeps into the descriptions offered by scientists. Other types of attractors are well-known, but these newly discovered ones were named *strange* by two scientists, David Ruelle and Floris Takens, because they wanted a name that was deeply suggestive (Gleick 1987, 131). As Ruelle says, "The name is beautiful and well-suited to these astonishing objects, of which we understand so little (in Coveney and Highfield 1990, 204).

To describe this dance between turbulence and order, Ruelle reaches for several metaphors: "These systems of curves, these clouds of points, suggest sometimes fireworks or galaxies, sometimes strange and disquieting vegetal proliferations. A realm lies there of forms to explore, of harmonies to discover" (in Coveney and Highfield 1990, 206). Briggs and Peat paint a similarly compelling picture of the drama and beauty of strange attractors forming: "Wandering into certain bands, a system is . . . dragged toward disintegration, transformation, and chaos. Inside other bands, systems cycle dynamically, maintaining their shapes for long periods of time. But eventually all orderly systems will feel the wild, seductive pull of the strange chaotic attractor" (1989, 76–77).

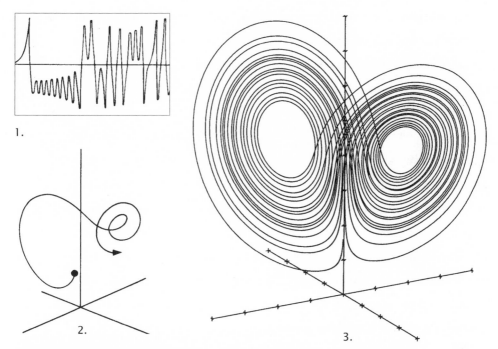

Strange Attractor. 1. Traditional plots of one variable show a system in chaos. 2. If the system is plotted in multiple dimensions in phase space, the shape of chaos, the strange attractor, gradually becomes visible. 3. As the system's chaotic wanderings are plotted over time (the system never repeats its behavior exactly), the attractor reveals itself. This butterfly or owl-shaped strange attractor reveals the order inherent in a chaotic system. Order always is displayed as a shape or pattern. *From Gleick, 1987. Used with permission.*

Chaos has always partnered with order—a concept that contradicts our common definition of chaos—but until we could see it with computers, we saw only turbulence, energy without predictable form. Chaos is the last state before a system plunges into random behavior where no order exists. Not all systems move into chaos, but if a system becomes unstable, it will move first into a period of oscillation, swinging back and forth between two different states. After this oscillating stage, the next state is chaos, and it is then that the wild gyrations begin. However, in the realm of chaos, where everything should fall apart, the strange attractor emerges, and we observe order, not chaos.

A strange attractor becomes visible on a computer screen because scientists have developed new ways of observing the system's wild and rich behavior. Its behavior is displayed in an abstract mathematical space called phase space. In phase space, scientists can track a system's movement in many more dimensions than was previously possible. Shapes that could not be seen in only two dimensions now appear, dancing on the screen, luminous and enticing.

In phase space, the system operates within a basin of attraction. This figurative basin is where the system explores millions of possibilities, wandering to different places, sampling new configurations of itself. But its wandering and experimentation respect a hidden boundary which is gradually revealed as the shape of its strange attractor. The system does not wander off into infinity. It is important to note that this boundary is not defined *for* the system; scientists do not create it. The boundary *lives within the system,* becoming visible as it explores its space of possibilities. The order is already present; it has now become discernible.

To see how chaotic processes reveal the order inherent in a system requires that we shift our vision from the parts to the whole. Briggs and Peat, in their exploration of the mirror world of chaos and order, suggest that wholeness is "what rushes in under the guise of chaos whenever scientists try to separate and measure dynamical systems as if they were composed of parts . . ." (1989, 74–75). The strange attractors that form on our screens, Briggs and Peat suggest, are not the shape of chaos. They are the shape of wholeness. When we concentrate on individual moments or fragments of experience, we see only chaos. But if we stand back and look at what is taking shape, we see order. Order always displays itself as patterns that develop over time.

In much of new science, we are challenged by paradoxical concepts— matter that is immaterial, disequilibrium that leads to stability, and now chaos that is ordered. Yet the paradox of chaos and order is not new. As ancient myths and new science both teach, every system that seeks to stay alive must hold within it the potential for chaos, "a creature slumbering deep inside the perfectly ordered system" (Briggs and Peat 1989, 62). It is chaos' great destructive energy that dissolves the past and gives us the gift of a new future. It releases us from the imprisoning patterns of the past by offering us its wild ride into newness. Only chaos creates the abyss in which we can recreate ourselves.

Most of us have experienced this ride of chaos in our own lives. At the personal level, chaos has gone by many names, including "dark night of the soul" or "depression." Always, the experience is a profound loss of meaning—nothing makes sense in the way it did before; nothing seems to hold the same value as it once did. These dark nights have been well-documented in many spiritual traditions and cultures. They are part of the human experience, how we participate in the spiral dance of form, formlessness, and new form. As we reflect on the times when we personally have descended into chaos, we can notice that as it ends, we emerge changed, stronger in some ways, new. We have held in us the dance of creation and learned that growth always requires passage through the fearful realms of disintegration.

Chaos' role in the emergence of new order is so well-known that it seems strange that Western culture has denied it's part so vehemently. In the dream of dominion over all nature, we believed we could eliminate chaos from life. We believed there were straight lines to the top. If we set a goal or claimed a vision, we *would* get there, never looking back, never forced to descend into confusion or despair. These beliefs led us far from life, far from the processes by which newness is created. And it is only now, as modern life grows ever more turbulent and control slips away, that we are willing again to contemplate chaos (see Hayles 1990). Whether we explore its dynamics through new science or ancient myths, the lesson is important. The destruction created by chaos is necessary for the creation of anything new.

Chaos theory studies a particular variety of chaos, known as deterministic chaos. In an interesting way, this branch of science became involved in a debate that had been going on in philosophy and spiritual thought for many centuries. Is this a deterministic world where our lives are predetermined? But if this is true, what about free will? It was this unresolved tension between predictability and freedom that attracted some early scientists of chaos. The science seemed to resolve this argument; it provided an explanation for how freedom functions in an orderly universe. The shape of the entire system is predictable or predetermined. But how this shape takes form is through individual acts of free agency: "The system is deterministic, but you can't say what it's going to do next" (Gleick 1987, 251). Or as organizational planner T. J. Cartwright puts it, "Chaos is order without predictability" (1991, 44).

The shape of chaos materializes from information feeding back on itself and changing in the process. This is the familiar process of iteration and feedback described in much of new science. It is the same process that results in self-organization, and also the creation of fractals (as noted in preceding chapters). This process succeeds in creating newness because it takes place in a system that is nonlinear. Nonlinearity has been described by Coveney and Highfield as "getting more than you bargained for" (1990, 184). In the past, science tended to ignore nonlinearity because it was just too hard to deal with. Science was focused on prediction, and nonlinear systems refuse prediction. To avoid the messiness and pursue the dream of determinism, nonlinear equations were "linearized." Once they were warped in this way, they could be handled by simpler mathematics. But this process of linearizing nature's nonlinear character blinded scientists to life's processes. Life, in the words of scientist Ian Stewart, is "relentlessly non linear." The recognition of nonlinearity and the newer mathematical tools of chaos theory have made it possible once again to see more clearly how life works (Capra 1996, Ch. 6).

In a nonlinear world, very slight variances, things so small as to be indiscernible, can amplify into completely unexpected results. When a system is nonlinear and webbed with feedback loops, repetition feeds the change back on itself, causing it to amplify and grow. After several iterations, a variance that was too small to notice can cause enormous impact, far beyond anything predicted. The system suddenly takes off in unexpected directions or responds in surprising ways. One familiar example of this is the proverbial straw that broke the camel's back. No one knew that such a small difference would cause collapse because no one could see what else had been going on inside the camel. In a nonlinear world, there is no relation between the strength of the cause and the consequence of the effect.

From classical science, our culture has come to believe that small differences average out, that slight variances converge toward a point, and that approximations can give a fairly accurate picture of what might happen. But chaos theory exposes the world's nonlinear dynamics, which in no way resemble the neat charts and figures we have drawn so skillfully. In a nonlinear system, the *slightest* variation can lead to catastrophic results. Hypothetically, were we to create a difference in two values as small as rounding them off to the thirty-first decimal place (calculating numbers this large requires astronomical computing power), after only one hundred iterations the whole calculation would go askew. The two systems would have diverged from each other in unpredictable ways. This behavior demonstrates that even infinitesimal differences can be far from inconsequential. "Chaos takes them," physicist James Crutchfield says, "and blows them up in your face" (in Briggs and Peat 1989, 73).

Edward Lorenz, a meteorologist, first drew public attention to this with his now famous "butterfly effect." Does the flap of a butterfly wing in Tokyo, Lorenz queried, affect a tornado in Texas (or a thunderstorm in New York)? Though unfortunate for the future of accurate weather prediction, his answer was "yes." And in organizations, we frequently experience these "flaps." A casual comment at a meeting flies through the organization, growing and mutating into a huge misunderstanding that requires enormous time and energy to resolve. And many organizations have learned that events occurring in a relatively minor part of their business suddenly grow to threaten their overall viability. Before disaster struck in Union Carbide's plant in Bhopal, India, the plant contributed a mere 4% to corporate profits. However, this horrific tragedy led to a major restructuring of the entire company and a serious decrease in its overall valuation. And in Alaska, how much ecological and cultural devastation on a grand scale was created from the actions of one oil tanker, the *Exxon Valdez*?

Science has been profoundly affected by this new relationship with the non-linear nature of our world. Many of the prevailing assumptions of scientific thought have had to be recanted. As scientist Arthur Winfree expresses it, the old dream of science was of a universe that was unaffected by slight changes:

> The basic idea of Western science is that you don't have to take into account the falling of a leaf on some planet in another galaxy when you're trying to account for the motion of a billiard ball on a pool table on earth. Very small influences can be neglected. There's a convergence in the way things work, and arbitrarily small influences don't blow up to have arbitrarily large effects (in Gleick 1987, 15).

But chaos theory has proved these assumptions false. The world is far more sensitive than we had ever dreamed. We may harbor the hope that we will regain predictability as soon as we can learn how to account for all variables. (Titles of conferences and books reveal this dream; two recent ones to cross my desk are "Conquering Uncertainty" and "Mastering Complexity.") But in fact these desires for mastery and prediction can never be satisfied in this non linear world. We would do better to abandon that search entirely. In nonlinear systems, iteration helps small differences grow into powerful and unpredictable effects. In complex ways that no model will ever capture, the system feeds back on itself, magnifying slight variances, communicating throughout its networks, becoming disturbed and unstable—and prohibiting prediction, ever.

Iteration launches a system on a journey that visits both chaos and order. The most beautiful consequences of iteration are found in the artistry of fractals. There is a difference between fractals and strange attractors. Strange attractors are self-portraits drawn by a chaotic system. They are always fractal in nature, being deeply patterned, but they are a special category of mathematical object. Estimates are that there are only about two dozen different strange attractors. In contrast, fractals describe any object or form created from repeating patterns evident at many levels of scale. There are an infinite number of fractals, both natural and human-made.

Fractals can be generated with computers by taking a few nonlinear equations and continuously feeding back into the system the results of those equations (see also Chapter 6). It is not any one solution that matters, but the composite picture of those behaviors that emerges after countless iterations. As individual solutions are plotted, the whole of the system emerges in the form of detailed, repetitive shapes.

Everywhere in this intricate fractal landscape, there is self-similarity. The shape we see at one magnification will be similar to what we'll find at all others. No matter how deeply we look, peering down through magnifications of more than a billion, the same forms are evident. There is pattern within pattern within pattern. There is no end to them, no scale small enough that these intricate shapes cease to form. We could follow the creation of these shapes forever, and at ever finer levels, there would always be something more to see (see color pages).

Fractals entered our world through the research of Benoit Mandelbrot, then at IBM. (Infinite patterns had been described in the early twentieth century by a few mathematicians, but their work lay dormant until quite recently.) In naming them, Mandelbrot gave us a language, a form of geometry, that allows us to comprehend nature in new ways. Fractals are everywhere around us, in

Broccoli's fractal qualities are easy to notice. The same shape appears at many different levels, from full head to tiny floret.

the patterns by which nature organizes clouds, rivers, mountains, many plants, tribal villages, our brains, lungs, and circulatory systems. All of these (and millions more) are fractal, replicating a dominant pattern at several smaller levels of scale (see color section). We live in a universe of fractal forms, but until recently, we lacked a means for seeing them. Now that we can see them, there are some wonderful lessons to be learned.

One primary lesson I have learned from fractals is that a world ordered by patterns does not explain itself through traditional measures. The infinite intricacy of fractals defies precise measurement. Mandelbrot's seminal fractal exercise was a simple question posed to colleagues and students: "How long is the coast of Britain?" As his colleagues soon learned, there is no answer to this question. As we zoom in, there are more and more details to measure. Creeping along the coastline, even if we chose to measure every rock on every outcrop, there would always be more to measure at ever smaller levels of scale.

Since fractals resist definitive assessment by familiar tools, they require a new approach to observation and measurement. What is important in a fractal

Three-winged Bird: A Chaotic Strange Attractor
This is a self-portrait drawn by a chaotic system.
The system's behavior is plotted over millions of
iterations. The system appears to be wandering
chaotically, always displaying new and different
behavior. But over time, a deeper order—a shape—
is revealed. This order is inherent to the system.
It was always there, but not revealed until its
chaotic movements were plotted in multiple
dimensions over time.

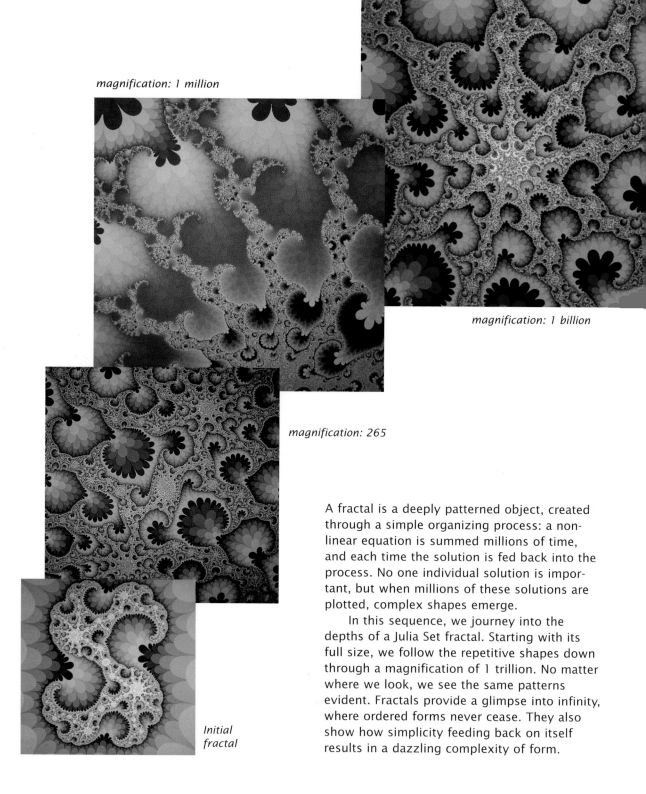

magnification: 1 million

magnification: 1 billion

magnification: 265

Initial fractal

A fractal is a deeply patterned object, created through a simple organizing process: a non-linear equation is summed millions of time, and each time the solution is fed back into the process. No one individual solution is important, but when millions of these solutions are plotted, complex shapes emerge.

In this sequence, we journey into the depths of a Julia Set fractal. Starting with its full size, we follow the repetitive shapes down through a magnification of 1 trillion. No matter where we look, we see the same patterns evident. Fractals provide a glimpse into infinity, where ordered forms never cease. They also show how simplicity feeding back on itself results in a dazzling complexity of form.

"Within its deep infinity
I saw ingathered,
and bound by love
in one volume,
the scattered leaves
of all the
universe."

—*Dante*

magnification: 40 billion

magnification: 1 trillion

It is fascinating to explore the fractal nature —the recurring patterns— of cumulus clouds from an airplane window.

In this scene of the Grand Canyon, other smaller canyons are evident as foreground. Photographers often capture the fractal qualities of nature, where repetitive patterns are easily evident at different levels of scale.

Computer-generated

Because of a natural fern's fractal nature, it is possible to create rich artificial ferns on computers. (see Chaos Game, p. 127)

Generated by Nature

Spiral patterns,
found in all nature and human art,
display the dance of order and chaos.

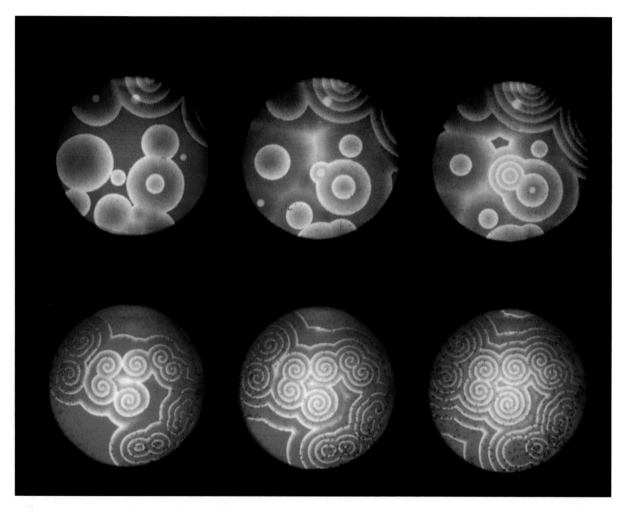

Belousov-Zhabotinsky Reaction
As a certain chemical mixture undergoes change, it self-organizes into
shapes far more complex than the original mixture. More intricate spirals
emerge as the change process continues.

A massive energy system self-organizes into a complex spiral. Many galaxies exhibit the same form.

Copper Double Spiral Ornament, Prehistoric
The spiral appears in human art all over the world, beginning with the Paleolithic period. Carl Jung believes the spiral is an archetype in the human psyche of the dance of creation and destruction.

Aurora Borealis
When the solar wind enters Earth's atmosphere, its charged particles stream to the electromagnetic poles. As the particles interact with nitrogen and oxygen, they become visible as colored light. These aurora demonstrate that space is not empty.

landscape is to note not quantity but *quality*. How complex is the system? What are its distinguishing shapes? How do its patterns differ from those of other systems? In a fractal world, if we ignore qualitative factors and focus on quantitative measures, we doom ourselves only to frustration. Instead of gaining clarity, our search for quantification leads us into infinite fogginess. The information never ends, it is never complete, we accumulate more and more but understand less and less. When we study the individual parts or try to understand the system through discrete quantities, we get lost. Deep inside the details, we cannot see the whole. Yet to understand and work with the system, we need to be able to observe it *as a system,* in its wholeness. Wholeness is revealed only as shapes, not facts. Systems reveal themselves as patterns, not as isolated incidents or data points (see Capra 1996, Ch. 3).

In organizations, we are very good at measuring activity. In fact, that is primarily what we do. Fractals suggest the futility of searching for ever finer measures that concentrate on separate parts of the system. There is never a satisfying end to this reductionist search, never an end point where we finally know everything about even that one small part of the system. Scientists of chaos study shapes in motion. If we were to understand organizations in a similar way, what would constitute the shapes in motion of an organization?

Different answers to this question are emerging from studies of organizations as whole systems. Learning to look for wholeness is a new skill for us, and it has been difficult not to rely on old measures, even when we know they don't give us the information we need. But seeing patterns is not a foreign skill for us; we are, after all, a pattern-recognizing species, and even as infants we are very adept at noticing them. But after so many years of data analysis that has left us drowning in increasing minutia, we need to help one another to reconnect with this innate ability. Together we must discipline ourselves to lift our heads from the pages and screens where charts dance hypnotically before us and enter into the world of form and shape.

The first step is to realize what we are looking for. A pattern has been defined rather succinctly as any behavior that occurs more than once. This seems elementary, but it is important to note what we are trying to see. So first we need to encourage each other to look for recurring behaviors and themes, to stay away from the seduction of examining isolated factors or individual players. Often patterns become discernible if we ask simple questions: "Have we seen this before?" "What feels familiar here?" To see patterns, we have to step back from the problem and gain perspective. Shapes are not discerned from close range. They require distance and time to show themselves. Pattern recognition requires that we sit together reflectively and patiently. I say patiently not only because patterns take time to form, but because we are trying to see the world differently and there are many years of blindness to overcome.

Fractals are extraordinarily complex objects. Their complex structure— such as the folds of human brains or the dense structure of lungs—provides increased capacity to process information and resources. But this complexity is created through processes that are quite different from human-created complexity. Fractal complexity originates in simplicity. Chaos scientist Michael Barnsley was intrigued to see if he could recreate the shapes of natural objects by deducing the simple equations that would describe their forms. He calls this the "Chaos Game." The game begins by ascertaining the essential information about the basic shape of the fractal (his first attempt was with a fern). These equations are surprisingly simple, devoid of the levels of precise prescriptive information we might think was necessary. He then sets the equations in motion to feed back on themselves. They are free to follow their own iterative wanderings, working at many different levels of scale, showing up in different sizes. With this approach, he can successfully reproduce an entire garden of plants on his computer (see color section).

The Chaos Game

What is the essential shape of
a complex, curvy fern?
It's a pattern of four straight lines.

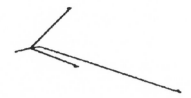

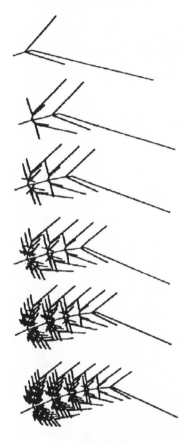

When this pattern repeats and repeats, free
to change size but not shape, the complexity
and beauty of the fern emerges. The pattern
must always connect with what is already on
the page, and in this example, it must
appear in an upright position.

All fractal patterns are created as
individuals exercise both freedom and
responsibility to a few simple rules.
Complex structures emerge over time from
simple elements and rules, and autonomous
interactions.

Drawing used with permission, Linda Garcia, 1991 (in
The Fractal Explorer, Dynamic Press, Santa Cruz, CA).

His work with fractals and the Chaos Game is surprising and instructive. First, Barnsley shows us that determinism still operates in this universe. The shapes that he creates are predictable, determined by the initial formula. But indeterminism also plays a key role. He cannot predict how the formula will next solve itself, or where the pattern will show up on the screen. It seems that with a few simple principles or formulas, combined with the freedom to develop and move about, nature creates the complexity and intricacy of form we see everywhere.

Many disciplines have seized upon fractals, testing whether self-similar phenomena occur at different levels of scale in both natural and human-made systems. From business forecasters and stock analysts who have observed a fractal quality to stock market behaviors, to physiologists who describe how the fractal quality of brain and lung tissue gives it far greater capacity, to architects who explain the beauty of buildings and towns as the repetition of harmonious patterns, fractals have entered the imagination and research of many disciplines. They have provided a very different lens for comprehending the workings of the natural world. They have revealed the partnering of chaos and order that gives birth to beauty.

And I believe that fractals have direct application for how we understand organizations. All organizations are fractal in nature. I can't think of any organization that isn't deeply patterned with self-similar behaviors evident everywhere. I am often struck by eerily similar behaviors exhibited by people in an organization, whether I'm meeting with a factory floor employee or a senior executive. I might detect a recurring penchant for secrecy, or for openness, for name-calling, or for thoughtfulness. These recurring patterns of behavior are what many call the culture of the organization. I believe we all experience this fractal nature of organizations in any of our encounters with them. As customers, we can learn how employees are treated by their bosses by noticing how the employees treat us. As a consultant, I was taught that I would be able to spot the dominant issues of the client system by noticing how the client interacted with me.

Fractal order originates when a simple formula is fed back on itself in a complex network. Except for the shape that is contained within this simple formula, there are no other constraints on behavior. Organizations that display a strong commitment to their values make good use of this fractal creation process. In these organizations, it doesn't matter where you go, whom you talk with, or what that person's role is. By observing the behavior of a production floor employee or a senior executive, you can tell what the organization values and how it chooses to do its work. You hear the values referred to even in casual conversation. You feel the values are real and alive. And in true fractal fashion, these vital agreements do not restrict individuals from embodying them in diverse and unique ways. Self-similarity is achieved not through compliance to an exhausting set of standards and rules, but from a few simple principles that everyone is accountable for, operating in a condition of individual freedom.

The potent force that shapes behavior in these organizations and in all natural systems is the combination of simply expressed expectations of purpose, intent, and values, and the freedom for responsible individuals to make sense of these in their own way. Organizations with integrity have truly learned that there is no choice but to walk their talk. Their values are truthful representations of how they want to conduct themselves, and everyone feels deeply accountable to them. Just as in the Chaos Game, the organization's principles contain sufficient information about the intended "shape" of the organization, what it hopes to accomplish and how it hopes to behave. When each person is trusted to work freely with those principles, to interpret them, learn from them, talk about them, then through many iterations a pattern of ethical behavior emerges. It is recognizable in everyone, no matter where they sit or what they do.

It is the nature of life to organize into patterns. This recognition welcomes us into a different approach to organizational change. We can see that it is important to look for and identify the patterns that reveal themselves through behavior. Together we can decide whether we would prefer different behaviors. If we do, we need to figure out the values and agreements that we think will support these new behaviors. Then we work together to see what it means to live into these new agreements. This work requires awareness, patience, and generosity. Behaviors don't change just by announcing new values. We move only gradually into being able to act congruently with those values. To do this, we have to develop much greater awareness of how we're acting; we have to become far more self-reflective than normal. And we have to help one another notice when we fall back into old behaviors. We will all slip back into the past—that is unavoidable—but when this happens, we agree to counsel one another with a generous spirit. Little by little, tested by events and crises, we learn how to enact these new values. We develop different patterns of behavior. We slowly become who we said we wanted to be.

These ideas speak with a simple clarity to issues of effective leadership. They recall us to the power of simple governing principles: guiding visions, sincere values, organizational beliefs—the few self-referential ideas individuals can use to shape their own behavior. The leader's task is first to embody these principles, and then to help the organization become the standard it has declared for itself. This work of leaders cannot be reversed, or either step ignored. In organizations where leaders do not practice what they preach, there are terrible disabling consequences. Barbara Ley Toffler, a consultant specializing in ethics, reports that employees respond with "less commitment to the institution, less commitment to the institution's goals, customers, and clients." She comments that senior executives "have got to really, genuinely, walk the talk, practice what they preach, live out what they say" (in McLenahen, 1999).

Leaders are also obligated to help the whole organization look at itself, to be reflective and learningful about its activities and decisions. Mort Meyerson, a retired CEO, says that one of the primary tasks of a leader is to make sure the organization knows itself (in "Everything I Knew About Leadership Is Wrong" 1996). The leader's role is not to make sure that people know exactly what to do and when to do it. Instead, leaders need to ensure that there is strong and evolving clarity about who the organization is. When this clear identity is available, it serves every member of the organization. Even in chaotic circumstances, individuals can make congruent decisions. Turbulence will not cause the organization to dissolve into incoherence.

When chaos has banged down the door and is tossing us around the room, it is difficult to believe that clear principles are sufficient. Anytime we experience chaos, our training urges us to interfere immediately, to rush in, to stabilize, to prevent further dissolution. Certainly one of the strongest critiques we make of each other is to say, "You're out of control." But if we can trust the workings of the world, we will see that the strength of our organizations is maintained if we retain clarity about the purpose and direction of the organization. When things become chaotic, this clarity keeps us on course. We are still able to make sense, even if the world grows mad.

In this chaotic world, we need leaders. But we don't need bosses. We need leaders to help us develop the clear identity that lights the dark moments of confusion. We need leaders to support us as we learn how to live by our values. We need leaders to understand that we are best controlled by concepts that invite our participation, not policies and procedures that curtail our contribution. During the past several years, there has been enough research to demonstrate the enduring strength and resiliency of companies that have strong values (Collins 2001, Collins and Porras 1993). But to this research we can now add the voice of chaos theory. Seemingly chaotic processes work with simple formulas to create astonishing complexity and capacity.

In chaos theory it is true that you can never tell where the system is headed until you've observed it over time. Order emerges, but it doesn't materialize instantly. This is also true for organizations, and this is a great challenge in our speed-crazed world. It takes time to see that a well-centered organization really has enough invisible structure to work well. Many of these organizations are already out there, beckoning to us from the future. But if they have not been part of our own experience, we are back to acts of faith. As the universe keeps revealing more of its ordering processes, hopefully we will understand that systems achieve order from clear centers rather than imposed restraints.

One of the mysteries of chaos theory is that no one knows where order comes from. Scientists don't design order into the initial equations. Ever since my imagination was captured by the phrase "strange attractor," I have contemplated whether such an organizing mystery exists in organizations. What is it that would be so attractive that it would hold our behavior within a boundary and keep us from wandering into formlessness? It seems clear to me now that values create such attractors. But by far the most powerful force of attraction in organizations and in our individual lives is *meaning*. Our greatest motivation in life, writes Viktor Frankl in his stunning presentation of logotherapy, "is not to gain pleasure or to avoid pain but rather to see a meaning . . ." (1959, 115).

In all types of organizations, too many filled with people exhausted, cynical, and burned-out, I have witnessed the incredible levels of energy and passion that can be evoked when leaders or colleagues take the time to recall people to the meaning of their work. It only takes a simple but powerful question: "What called you here? What were you dreaming you might accomplish when you first came to work here?" This question always elicits a deep response because so few of us work for trivial purposes. Most people come to their organizations with a desire to do something meaningful, to contribute and serve. Everybody needs, as philosopher and management scholar Charles Handy says, "an inner belief that you are in some sense meant to be here, that you can leave the world a little different in a small way" (in Hesselbein and Cohen 1999, 130). If we are asked to recall that inner belief, and if we hear our colleagues speak about their own yearnings to make a small difference, we feel new energy for the work and for each other. The call of meaning is unlike any other, and we would do well to spend more time together listening for the deep wells of purpose that nourish all of us.

One quality particular to human beings is the need to know "Why?" We need to understand and ascribe meaning to things. When we are able to reflect on our experience and develop our interpretation, we can endure even the most horrendous events. Even horrific accidents do not appear then as random assaults; we make sense of them from a grander logic. As organizations continue to experience so many momentous challenges, we do a great disservice to one another if we try to get through these times by staying at a superficial level or believing we are motivated only by self-interest. We have a great need to understand from a larger perspective why we are confronted with dislocation and loss. We have to be willing to speak about events from this deeper level of meaning.

We also need to acknowledge the difficult side of life—the sorrow and suffering that has come into our experience. We surface these dark shadows not to mend them or make them disappear, but simply to acknowledge they are part of the reality of life. When leaders honor us with opportunities to know the truth of what is occurring and support us to explore the deeper meaning of events, we instinctively reach out to them. Those who help us center our work in a deeper purpose are leaders we cherish, and to whom we return love, gift for gift. It is only meaning that enables us to summon our Gaian energy from Chaos' depths. With meaning as our centering place, we can journey through

Spirals etched into a temple stone around 3000 B.C. in Malta. These spirals sprout leaves, depicting how chaos gives birth to new life.

the realms of chaos and make sense of the world. With meaning as our attractor, we can recreate ourselves to carry forward what we value most.

We can use our own lives as evidence for this human thirst for meaning. As we mature in life, we search to see a deeper and more coherent purpose behind the events and crises that compose our lives. What shape has my life taken? What is my purpose? Can I now see that seemingly random events were part of a greater plan? Do "chance" meetings now seem to have been not at all accidental? Each of us seeks to discover a meaning to our life that is wholly and uniquely our own. We experience a deepening confidence that purpose has shaped our lives, even as it moved invisibly in us. Whether we believe that we create this meaning for ourselves in a senseless world, or that it is offered to us by a purposeful universe, it is, after all, only meaning that we seek. Nothing else is attractive; nothing else has the power to cohere an entire lifetime of activity. We become like ancient Gaia, boldly embracing the void, knowing that out of Chaos' dark depths we have the strength to give birth to order.

Each time a person stands up for an idea, or acts to improve the lot of others, or strikes out against injustice, (s)he sends forth a tiny ripple of hope, and crossing each other from a million different centers of energy and daring, those ripples build a current that can sweep down the mightiest walls of oppression and resistance.

— Robert F. Kennedy

Chapter 8

Change: The Capacity of Life

We live in a time of great stirring storms, both natural and human-made. Disruptive elements seem to be afoot, gathering strength in air masses that spiral over oceans or in decisions that swirl through the halls of power. The daily news is filled with powerful changes, and many of us feel buffeted by forces we cannot control. It was from this place of feeling battered and bruised that I listened one night to a radio interview with a geologist whose specialty was beaches and shorelines. The interview was being conducted as a huge hurricane was pounding the Outer Banks of the eastern United States. The geologist had studied the Outer Banks for many years and was speaking fondly about their unique geological features. He was waiting for the storm to abate so he could get out and take a look at the hurricane's impact. The interviewer asked: "What do you expect to find when you go out there?" Like the interviewer, I assumed he would present a litany of disasters— demolished homes, felled trees, eroded shoreline. But he surprised me. "I expect," he said calmly, "to find a new beach."

Since that night, I have pondered what it would take for me and my colleagues to bring his clarity to our own work, to understand that this world changes, to be curious about newness. We live in the same world as this geologist, but in the organizations that I work in, change is a feared enemy. Hurricanes, organizational crises, sudden accidents—these are terrible forces that can destroy the deliberate, incremental progress we're all working hard to achieve. We haven't thought that we might *work with the forces of change.* We act quite the opposite; we need to manage change and keep it under control every cautious step of the way. And we think we're being helpful to others when we manage change so carefully, because we believe that people don't like change. Strangely, we assert that it's a particular characteristic of the human species to resist change, even though we're surrounded by tens of millions of other species that demonstrate wonderful capacities to grow, adapt, and change.

Our ideas and sensibilities about change come from the world of Newton. We treat a problematic organization as if it was a machine that had broken down. We use reductionism to diagnose the problem; we expect to find a simple, singular cause for our woes. We sift through all the possible causes of failure, searching for that one broken part—a bad manager, a dysfunctional team, a poor business unit. To repair the organization, all we need to do is replace the faulty part and gear back up to operate at predetermined performance levels.

This is the standard approach to organizational change. It is derived from the best engineering thinking. I believe this approach explains why the majority of organizational change efforts fail. Senior corporate leaders report that up to 75% of their change projects do not yield the promised results. This is a shocking failure rate, but how can we expect anything better until we stop treating organizations as machines?

We also display the influence of Newton when we define the size and scope of our change projects. We think we need to develop sufficient mass to counteract the organization's material weight. In classical physics, mass is important. An object's force is equal to two factors, its mass and its acceleration. We act on this law; if we are trying to change a large organization, we either need large change projects, where the force of our efforts equals the organization's mass, or smaller projects that have a lot of speed. Whichever strategy we choose, we worry about how to influence the organization's *physical size*.

But when we encounter life's processes for change, we enter a new world. We move from billiard balls banging into one another to effect change, to networks that change because of information they find meaningful. We stop dealing with mass and work with energy. We discard mechanistic practices, and learn from the behavior of living systems. New change dynamics become evident.

The new sciences are filled with tantalizing and hopeful processes that foster change. But to learn these lessons, we need to shift what we look for. Many of the reformulations of new science came from just such a shift: Scientists learned to look past an object or thing to the invisible level of dynamic processes. Laying aside the machine metaphor, with its static mechanisms and separated parts, scientists saw something new. They saw the underlying processes that give rise to innumerable and different life forms. They developed answers to explain how life is capable of so much change, so much newness. Some expressed awe and humility as they encountered the unstoppable resiliency of life. Some became poets, reaching for new language to describe their encounters with life's boundless creativity.

I am hopeful that we non-scientists can now make a similar shift. Surrounded by creativity expressed as unending diversity, living in a world proficient at change, which maintains its resiliency through change, I hope we can begin to work with these powers rather than seeking to control or deny them. But the shifts required of us are enormous; they lead us into lands that are foreign and uncharted in Western thought.

The first great shift is this. A system is composed of parts, but we cannot understand a system by looking only at its parts. We need to *work with the whole of a system*, even as we work with individual parts or isolated problems. From a systems consciousness, we understand that no problem or behavior can be understood in isolation. We must account for dynamics operating in the whole system that are displaying themselves in these individual moments. In earlier chapters, I described what this new orientation revealed in both quantum physics and chaos theory. When scientists shifted their vision from the parts to the whole, what looked like chaos revealed inherent order; a chaotic system displayed itself in a strange attractor. What seemed like an aberration of Newtonian laws became lawful; paired electrons refused to act individually and exhibited their inseparable wholeness across vast distances. A systems world cannot be understood by looking only at discrete events or individuals.

But learning to observe the whole of a system is difficult. Our traditional analytic skills can't help us. Analysis narrows our field of awareness and actually prevents us from seeing the total system. We move deeper into the details and farther away from learning how to comprehend the system in its wholeness. Hans-Peter Dürr, former director of the Max Planck Institute, once remarked to me, "There is no analytic language to describe what we are seeing at the quantum level. I can only say that it does not help to analyze things in more detail. The more specific the information, the less relevant it is."

If we can't analyze wholeness, how then do we learn to know it? This is a question that has occupied philosophers and some scientists for many centuries. They each describe new ways of understanding, but their answers feel insufficient. They fail to provide the precise, analytic techniques we think we need to understand anything. I frequently get frustrated by the realization that to perceive the world differently requires new perceptual techniques. We can't move past analysis by being analytic. But if I can't use my traditional ways of knowing, how can I even know enough about a new phenomenon to acknowledge that I need new ways of knowing? (So if you feel frustrated by the following descriptions, I believe this indicates you're making progress.)

As I have struggled to understand a system as a system, I have been drawn to move past cognition into the realm of sensation. The German philosopher Martin Heidegger describes this as a "dwelling consciousness." When we dwell with a group or a problem, we move quietly into our senses, away from our sharpened analytic skills. Now I allow myself to pick up impressions, to notice how something feels, to sit with a group or with a report and call upon my intuition. I try to encourage myself and others to look for images, words, patterns that surface as we focus on an issue. (The Army has been aware that intuition plays a role in their effectiveness; a few years ago, they began studying "commander intuition.")

The great scientist, philosopher, and poet of the early nineteenth century, Johann von Goethe, applied his genius to the problem of seeing the wholeness of nature. He was intrigued to understand any phenomenon not as an isolated event, but as a consequence of its *relationship* to other phenomena. In traditional science, the scientist invents the questions and then interrogates the object of study. But Goethe describes how we can move from interrogation to receptivity, being open to what is occurring, allowing ourselves to be influenced by a whole that we cannot see. We can dwell with the phenomenon and feel how it makes itself known to us.

Goethe describes several ways to sense the whole, and I am particularly challenged by one of his processes—that we can discover the whole by going further into its parts. While this sounds like good old-fashioned reductionism, it is quite different. We inquire into the part as we hold the recognition that it is participating in a whole system. We hold our attention at two levels simultaneously. We recognize that this one thing we are studying is only there because of the rest of the universe (see Bortoft, 1996, 6). We can understand the whole by noting how it is influencing things at this local level. This manner of thinking, while difficult to grasp for a Western mind, is familiar in Buddhist belief, as illustrated in this brief teaching story:

All things depend on all other things for their existence. Take, for
example, this leaf. . . . Earth, water, heat, sea, tree, clouds, sun, time,
space—all these elements have enabled this leaf to come into existence. If
just one of these elements was missing, the leaf could not exist. All beings
rely on the law of dependent co-arising. The source of one thing is all
things. (in Thich Nhat Hanh, 1991, 169)

To study a problem from this sensibility requires us to explore the
relationship between the part and the whole, but not to confuse them as
identical or interchangeable. This is a different exploration than looking at a
system for its fractal patterns or holographic images; in that search, we would
look at the part as a miniature version of the whole. Instead, here we look
intently at the part in order to see the dynamics operating in the whole system.
The part is not the whole, but it can lead us there.

Mostly we don't take time to notice the dynamics that are moving in the
whole system, creating effects everywhere. As good engineers, we've been
trained to identify the problem part and replace it. But a systems sensibility
quickly explains why this repair approach most often fails. Individual behaviors
co-evolve as individuals interact with system dynamics. If we want to change
individual or local behaviors, we have to tune into these system-wide
influences. We have to use what is going on in the whole system to understand
individual behavior, and we have to inquire into individual behavior to learn
about the whole.

Although we've all been trained in reductionist modes of analysis, many
people in organizations know firsthand that studying problems in detailed
isolation doesn't yield the promised improvements and changes. When I've
asked "If we were to solve all the individual problems, every one of them,
would this fix the organization?" most people reply "No." Clearly, they
understand that there are other forces at work, holding the organization in its

troubled state. They may not be able to name them, but they know that they're there.

Seeing the interplay between system dynamics and individuals is a dance of discovery that requires several iterations between the whole and its parts. We expand our vision to see the whole, then narrow our gaze to peer intently into individual moments. With each iteration, we see more of the whole, and gain new understandings about individual elements. We paint a portrait of the whole, surfacing as much detail as possible. Then we inquire into a few pivotal events or decisions, and search for great detail there also. We keep dancing between the two levels, bringing the sensitivities and information gleaned from one level to help us understand the other. If we hold awareness of the whole as we study the part, and understand the part in its relationship to the whole, profound new insights become available.

There are many processes for developing awareness of a whole system—a time line of some slice of the system's history, a mind-map, a collage of images, a dramatization. Any process works that encourages nonlinear thinking and intuition, and uses alternative forms of expression such as drama, art, stories, and pictures. The critical task is to evoke our senses, not just our gray matter. We learn to dwell in multilevel phenomena simultaneously and let our senses lead us to new ways of comprehending.

In one corporation, a business unit wanted to know why they failed to secure a major contract. First, they developed a time line of all the events and decisions they could recall. Everybody had to participate; no one person knew the whole story. (The time line ended up being more than thirty feet long.) Everyone reviewed it, developing a rudimentary sense of the whole system that had resulted in this business failure. Next, the whole group defined which decisions, among the many displayed, felt most critical. They then went into small groups, each group exploring in depth one of those decisions. But because they had started with the whole, their search to understand the parts was already different. Each group then brought its analysis of single decisions back to the whole time line. It became instantly clear that similar patterns of behavior characterized each of these decisions. The whole was revealing its dynamics in each event, but no one would have seen these patterns had they not been aware of the whole. After another iteration of going deeply into different parts of the experience and bringing these back to the whole, a few dynamics stood out clearly. The real work of change came into focus: how to shift those dynamics.

This kind of work must involve the whole group. The whole must go in pursuit of itself; there is no other way to learn who they are. But as people engage together to learn more about their collective identity, it affects them as individuals in a surprising way. They are able to see how their personal patterns and behaviors contribute to the whole. The surprise is that they then take responsibility for changing themselves.

It's important to note that the motivation for individual change is not in response to a boss's demand or a personal need for self-improvement. A larger context has emerged because of this collaborative process, and it is this context that motivates people to change. They have developed a deeper awareness of the work, not of personalities or particular parts of the organization. They want *the work* to be more effective, and they now see how they individually can better contribute to that outcome.

If the first shift challenges us to think differently about parts and wholes, the second shift focuses us on the organizing dynamics of a living system. The organization of a living system bears no resemblance to organization charts. Life uses networks; we still rely on boxes. But even as we draw our boxes, people are ignoring them and organizing as life does, through networks of relationships. To become effective at change, we must leave behind the *imaginary organization* we design and learn to work with the *real organization*, which will always be a dense network of interdependent relationships.

The new science keeps reminding us that in this participative universe, nothing living lives alone. Everything comes into form because of relationship. We are constantly called to be in relationship—to information, people, events, ideas, life. Even reality is created through our participation in relationships. We choose what to notice; we relate to certain things and ignore others. Through these chosen relationships, we co-create our world.

If we are interested in effecting change, it is crucial to remember that we are working with these webs of relations, not with machines. Once we recognize that organizations are webs, there is much we can learn about organizational change just from contemplating spider webs. Most of us have had the experience of touching a spider web, feeling its resiliency, noticing how slight pressure in one area jiggles the entire web. If a web breaks and needs repair, the spider doesn't cut out a piece, terminate it, or tear the entire web apart and reorganize it. *She reweaves it*, using the silken relationships that are already there, creating stronger connections across the weakened spaces.

The most profound strategy for changing a living network comes from biology, although we could learn it directly from a spider. If a system is in trouble, it can be restored to health by connecting it to more of itself. To make a system stronger, we need to create stronger relationships. This principle has taught me that I can have faith in the system. The system is capable of solving its own problems. The solutions the system needs are usually already present in it. If a system is suffering, this indicates that it lacks sufficient access to itself. It might be lacking information, it might have lost clarity about who it is, it might have troubled relationships, it might be ignoring those who have valuable insights.

To bring health to a system, connect it to more of itself. The primary change strategy becomes quite straightforward. In order to change, *the system needs to learn more about itself from itself*. The system needs processes to bring it together. Many different processes will work, whatever facilitates self-discovery and creates new relationships simultaneously. The whole system eventually must be involved in doing this work; it can't be done by outside experts or small teams.

My colleagues and I focus on helping a system develop greater self-knowledge in three critical areas. People need to be connected to the fundamental *identity* of the organization or community. Who are we? Who do we aspire to become? How shall we be together? And people need to be connected to *new information*. What else do we need to know? Where is this new information to be found? And people need to be able to reach past traditional boundaries and develop *relationships* with people anywhere in the system. Who else needs to be here to do this work with us?

As a system inquires into these three domains of identity, information, and relationships, it becomes more self-aware. It has become more connected to the truth of who it is, more connected to its environment and customers, more connected to people everywhere in the system. These new connections develop greater capacity; the system becomes healthier.

There are many stories of increased organizational effectiveness gained from creating new connections in these three domains, although frequently even the implementers seem not to understand the source of their success. For example, in the very best quality programs, employees were first connected to a new identity or meaning for their work, such as exceptional customer service or the design of highly productive work processes. New statistical tools gave these employees new information about their work. They could use this to achieve and often surpass the new standards they had set for themselves. Participative problem-solving processes and self-managed teams facilitated workers' connecting with one another and sharing expertise. Just as important were the new connections with customers and suppliers—those formerly estranged from the system were invited inside to contribute.

The novelist E. M. Forster said "Just connect." But of course, it's not quite that simple. We have all been to many events and meetings that offered great opportunity for people to connect with each other, yet nothing happened. People didn't step forward to find one another, nothing significant was discussed, everyone hid in their own boxes waiting to be coaxed out.

These bad parties and boring meetings illustrate the next shift to consider as we learn to work with life's capacity for change. Any living thing will change only if it *sees change as the means of preserving itself.*

All life lives in the midst of an unending stream of data. How do we select what to pay attention to from so much noise? We use the lens of self. We, like all life, choose what to notice because of who we are. We use the process of *self-reference.* We are free to choose, but we choose on the basis of self. This process is essential for all life and, if repressed or denied, the organism dies. Self-reference explains *why* any living system is motivated to change. It will change to stay the same.

In humans, self-reference becomes more complex because of capacities that differentiate us from most other species. We possess consciousness and are capable of reflection. We are able to think about a past and a future. No longer anchored to just the present moment, we can dream about what we want, and imbue events with meaning. We still see the world through a self, but to this self has been added the dimensions of time and meaning.

It's hard to look at modern life and see our capacities for reflection or meaning-making. We don't use our gifts to be more aware or thoughtful. We're driven in the opposite direction. Things move too fast for us to reflect, demanding tasks give us no time to think, and we barely notice the lack of meaning until forced to stand still by illness, tragedy, or job loss. But in spite of our hurry, we cannot stop life's dynamic of self-reference or the human need for meaning. If we want to influence any change, anywhere, we need to work *with* this powerful process rather than deny its existence. We need to understand that all change results from a change in meaning. Meaning is created by the process of self-reference. We change only if we decide that the change is meaningful to who we are. Will it help us become who we want to be? Or gain us more of what we think we need to preserve ourselves?

From becoming attuned to this dynamic, I've come to believe that both individual and organizational change start from the same place. People need to explore an issue sufficiently to *decide whether new meaning is available and desirable.* They will change only if they believe that a new insight, a new idea, or a new form helps them become more of who they are. If the work of change is at the level of an entire organization or community, then the search for new meaning must be done as a collective inquiry.

To put this realization into practice has required significant changes on my part. Now my first desire with a group is to learn who they are, what self they are referencing. I can never learn this by listening to some self-reports, or taking the word of a few people. I discover who they are by noticing what's meaningful to them as they are engaged in their work. What issues and behaviors get their attention? What topics generate the most energy, positive and negative? I have to be curious to discover these answers. And I have to be *working with them*, not sitting on the side observing behavior or interviewing individuals. In the process of doing actual work, the real identity of the group, not some fantasy image, always becomes visible.

There's another aspect to this work that is important to me. I assume that even in the presence of a group or collective identity, there are as many different interpretations as there are people in the group. I assume I will discover multiple and divergent interpretations for everything that occurs. So I try to put ideas and issues on the table as experiments to discover these different meanings, not as my recommendations for what *should* be meaningful. I try to stay open to the different reactions I get, rather than instantly categorizing people as resistors or allies (although this is not always easy). I expect diverse responses; gradually, I'm even learning to welcome them. It has been fascinating to notice how many interpretations the different members of a group can give to the same event. I am both astonished and confident that, as quantum theory and biology teach, no two people see the world exactly the same.

However you do it, discovering what is meaningful to a person, group, or organization is the first essential task. We discover this by looking into our actual, day-to-day work. It doesn't help to go off and talk about meaning or behaviors in the abstract. We need to be able to see what we are doing as we are doing it; this is where the true learning is. To develop this "observer self" requires practice, curiosity, and patience.

But as we engage in this process of exploring diverse interpretations and learning to observe our patterns, oftentimes we discover a unifying energy that makes the work of change possible. If we discover an issue whose significance we share with others, those others are transformed into colleagues. If we recognize a shared sense of injustice or a common dream, magical things happen to people. Past hurts and negative histories get left behind. People step forward to work together. We don't hang back, we don't withdraw, we don't wait to be enticed. We seek each other out, eager to discover who else might help. The call of the problem sounds louder than past grievances or our fears of failure. We have found something important to work on, and, because we want to make a difference, we figure out how to do the work, together.

I've come to appreciate that real change happens in personal behaviors, or at larger scale in entire organizations, only when we take time to discover this sense of what's worthy of our shared attention. We don't accept an organizational redesign because a leader tells us it is necessary. We choose to accept it if, and only if, we see how this new design enables us to contribute more to what we've defined as meaningful. And we don't accept diversity because we've been told it's the right thing to do. Only as we're engaged together in work that is meaningful do we learn to work through the differences and value them. Change becomes much easier when we focus first on creating a meaning for the work that can embrace us all. Held by this rich center of meaning, we let go of many other grievances and work around traditional hindrances.

I've worked with some college faculties torn apart by the availability of technology. The more technologically eager faculty accuse the reticent ones of being out-of-date and resistant to change—they berate their colleagues for not climbing on the technology bandwagon. I always suggest that a different conversation is needed. What if we stop assuming that technology's value to a teacher is self-evident? What if we stop assuming that anybody who doesn't adopt new technology is an antiquated Luddite whose only interest is to stop the march of progress? If we give up those assumptions, we can begin a different conversation, one that helps us connect to one another and learn more about how we each see the world. We can step back from the technology issue and ask one another what called us into teaching. We can listen to the aspirations that are voiced. What we will hear is that most of us went into teaching for noble purposes—we wanted to make a difference in the lives of students and to advance human wisdom.

If we have this conversation *first*, we can discover one another as colleagues. *Then* we are ready to talk about technology. How might computers assist a professor to become more effective at his or her craft? How might technology make it easier to do the work they have defined as meaningful? If those links are made between professional purpose and technical tools, more colleagues will log on to e-mail, and use the computers sitting on their desks.

This process of inquiring into the meaning of our work helps us move past the labeling behavior so common these days. We are quick to assign people to a typology and then dismiss them, as if we really knew who they were. And our frantic need to implement changes we know are crucial to our organization's survival leads us to grasp for scapegoats. We know we'd be successful if it weren't for all those "resistors," those stubborn and scared colleagues who reject anything new. (We label ourselves also, but more generously, as "early adopters" or "cultural creatives.")

In our crazed haste, we don't have time to be curious about who a person is, or why they're behaving as they do. But when we dwell in the meaning we each ascribe to our work, we might discover common issues and problems that we both deem significant. Then change becomes possible. We move past the labels and notice another human being who wants to make some small contribution to something we care about. We discard the divisive categories and want to work together. How else but through our joining can we create the change we both want to see in the world?

Meaningful information lights up a network and moves through it like a windswept brushfire. Meaningless information, in contrast, smolders at the gates until somebody dumps cold water on it. The capacity of a network to communicate with itself is truly awe inspiring; its transmission capability far surpasses any other mode of communication. But a living network will transmit only *what it decides is meaningful*. I have watched information move instantaneously across great distances in a global company; I have watched information in four-color graphics die before it ever came off the printer. To use a network's communication capacity, we must notice that its transmission power is directly linked to the meaningfulness of the information.

From witnessing how networks can communicate around the world with information they deem essential, I've come to believe that "preaching to the choir" is exactly the right thing to do. If I can help those who already share certain beliefs and dreams sing their song a little clearer, a little more confidently, I know they will take that song back to their networks. I don't have to touch everybody; I just have to support those first courageous voices and encourage them to put it out on their own airwaves. Soon large populations in diverse places will have heard the song because someone in their network had their voice amplified by meeting the choir. We gain courage from learning we're part of a choir. We sing better when we know we're not alone.

Nothing described by Newtonian physics has prepared us to work with the behavior of living networks. We were taught that change occurs in increments, one person at a time. We not only had to design the steps; we also had to take into account the size of the change object. The force of our efforts had to equal the weight of what we were attempting to change. But now we know something different. We're working with networks, not billiard balls. We don't have to push and pull a system, or bully it to change; we have to participate with colleagues in discovering what's important to us. Then we feed that into our different networks to see if our networks agree.

In working with networks, size is not the issue. The same fundamental dynamics are always at work in any living system, no matter how small or large. Self-reference and meaning-making never cease; therefore, change is always possible through those processes. Of course, people in different locations or levels of an organization will have interpretations and dynamics specific to them. But the work of change is always the same. We need to find ways to get their attention; we need to discover what's meaningful to them. Size doesn't matter, but meaning does.

As we contemplate how networks change themselves, it helps to remember that we are working with energy, not matter. Energy behaves differently than matter. It fills the universe, possibly traveling many times faster than the speed of light. It moves through invisible media and connections. Meaning has many of the qualities of energy. It doesn't exist in physical form anywhere. We make it up as we self-reference our way through life. Since it doesn't exist in material form, it too is not subject to the laws that govern matter. Its behavior can't be explained by Newtonian physics.

The energetic nature of meaning is another reason to give up organizational change strategies that are based on Newtonianism and the manipulation of discrete pieces. Matter doesn't matter. We can stop striving to achieve critical mass, we can let go of the need for programs that roll out (or over) the entire organization, we can abandon the need to train every individual, we can stop feeling thwarted if we don't get the support of the top of the organization. Instead, we can work locally, finding the meaning-rich ideas and processes that create energy in one area of the system. If we succeed in generating energy in one area, then we can watch what our networks do with our work. Who lit up and took notice? Where have our ideas traveled to? If we answer these questions, we learn who might be ready to take up this work next. My colleague, Myron Rogers, describes his approach to organizational change as "I start anywhere and follow it everywhere."

In this chapter, I've described what I believe to be the primary processes of life that facilitate change. If we are to ally ourselves with these processes and life's extraordinary capacity for change, there is one last essential shift in our thinking. Although we see change at the material level, it is caused by processes that are immaterial. We must look for these *invisible processes rather than the things that they engender.* From the early Greek Heraclitus to the most recent thinking in science, life is described as a process, a process of becoming (Prigogine 1998, 10). When scientists look behind the physical manifestations, or peer into the emptiness of space or cells, they see what had gone unnoticed —the processes that give rise to forms. Similar work is now required of us in organizations. We must look behind the things of organizations to work with the processes that gave them birth.

This shift in orientation requires many new practices, some of which I've indicated or described. But the greatest challenge for me lies not in adopting any one new method, but in learning generally to live in a process world. It's a completely new way to be. Life demands that I participate with things as they unfold, to expect to be surprised, to honor the mystery of it, and to see what emerges. These were difficult lessons to learn. I was well-trained to create things—plans, events, measures, programs. I invested more than half my life in trying to make the world conform to what I thought was best for it. It's not easy to give up the role of master creator and move into the dance of life.

But what is the alternative, for me or you? Our dance partner insists that we put ourselves in motion, that we learn to live with instability, chaos, change, and surprise. We can continue to stand immobilized on the shoreline, trying to protect ourselves from life's insistent gales, or we can begin moving. We can mourn the erosion of our plans, or we can set out to discover something new.

Morihei Ueshiba, the founder of the martial art of Aikido, was a small man who could turn back the onslaughts of opponents many times his weight and size with movements that were imperceptible. He appeared to be perfectly centered, anchored to the ground in an extraordinary way. But this was not the case. His ability came not from superior balance, but from superb levels of self-awareness. As he described it, he was quicker to notice when he was off-balance, and faster at returning to center.

He perfectly describes how to move in harmony with life rather than to resist it. First, we must know what "center" feels like. We must know who we are, our patterns of behavior, our values, our intentions. The ground of our identity and experience must feel familiar to us; we must know what it feels like to be standing in it. But we don't expect that we will be perfectly balanced in that center all the time. We know that we will drift into the wrong activities or be thrown off-balance by life's chaos. But we also will recognize when we've moved off too far, and will be able to recall ourselves more quickly to who we want to be.

Ueshiba Sensei also highlights a quality of attention—we must keep participating in the moment. The changing nature of life insists that we stop hiding behind our plans or measures and give more attention to what is occurring right in front of us, right now. We need to become curious about what's going on, what just happened. The present moment overflows with information about ourselves and our environment. But most of those learnings fly by unobserved because we're preoccupied with our images of how we want the world to be.

Being present in the moment doesn't mean that we act without intention or flow directionless through life without any plans. But we would do better to attend more carefully to *the process* by which we create our plans and intentions. We need to see these plans, standards, organization charts not as objects that we complete, but as processes that enable a group to keep clarifying its intent and strengthening its connections to new people and new information. We need less reverence for the objects we create, and much more attention to the processes we use to create them. Healthy processes create better relationships among us, more clarity about who we are, and more information about what's going on around us. With these new connections, we grow healthier. We develop greater capacity to know what to do. We weave together an organization as resilient and flexible as a spider's web.

As we learn to live and work in this process world, we are rewarded with other changes in our behavior. I believe we become gentler people. We become more curious about differences, more respectful of one another, more open to life's surprises. It's not that we become either more hopeful or pessimistic, but we do become more patient and accepting. I like to believe we change in this way because we are willing to move into the dance. Although it looked frantic from the outside, difficult to learn and impossible to master, our newfound gentleness speaks to a different learning. Life is a good partner. Its demands are not unreasonable. A great capacity for change lives in every one of us.

Science outstrips other modes & reveals
more of the crux of the matter
than we can calmly handle.

—A. R. Ammons

Chapter 9
The New Scientific Management

In the history of human thought, a new way of understanding often appears simultaneously in widely separated places and in different disciplines. These synchronicities, mysterious and inexplicable, pop up everywhere. For example, Darwin proposed his theory of evolution at the same time that another researcher, working in Malaysia, published very similar ideas. Physicist David Peat traces how the understanding of light evolved in parallel ways in both art and science over the centuries, a relationship that continues to this day. The sixteenth-century Dutch school of painters drew light for its effects on interior spaces, depicting how it entered rooms through cracks or under doors or was transformed as it passed through colored glass. At the same time, Sir Isaac Newton was studying prisms and the behavior of light as it passed through small apertures. Two hundred years later, the English landscape artist J. M. W. Turner painted light as energy, a swirling power that dissolved into many forms; simultaneously, physicist James C. Maxwell was formulating his wave theory in which light results from the swirling motion of electrical and magnetic fields. When Impressionist painters explored light for its effects on dissolving forms, even painting it as discrete dots, physicists were theorizing that light was made up of minuscule energy packets known as quanta (Peat 1987, 31–32; Schlain 1991).

We live again in a time when the same concepts are appearing in many places. These concepts are our *zeitgeist*—a way of thinking that characterizes a generation or time period. Our zeitgeist is a new (and ancient) awareness that we participate in a world of exquisite interconnectedness. We are learning to see systems rather than isolated parts and players. Under rather austere titles of *systems thinking* or *ecological thinking,* we are discovering many things worthy of wonder. We can now see the webs of interconnections that weave the world together; we are more aware that we live in relationship, connected to everything else; we are learning that profoundly different processes explain how living systems emerge and change. Many disciplines, in different voices, now speak about the behavior of networks, the primacy of relationships, the importance of context, and new ways to honor and work with the wholeness of life.

These parallel concepts are quite evident in both science and business. The world of electronic networks and connectivity that we depend upon mirrors the images from quantum physics that describe our interconnectedness at the cosmic scale. Scientists and businesspeople use surprisingly similar language to describe this new world. When Levi Strauss' former CEO Robert Haas describes today's world of business, he says that "we are at the center of a seamless web of mutual responsibility and collaboration, a seamless partnership, with interrelationships and mutual commitments." It is easy to hear a similar sensibility in the voices of scientists (in Howard 1990, 136).

Another parallel path being walked by both business and science is the recent work to understand living systems. Some organizational theorists and leaders are drawing on insights emerging from ecology, biology, and evolutionary theory. We look hopefully to nature to teach us how to do what living systems accomplish with such skill—learn, adapt, and change. Our interest is prompted by the relentless need for organizations to grow and re-form at intervals so short that change has become a continuous demand. We speak about "organic" organizations, self-organization, and emergent properties. Others are attracted to chaos and complexity theory in hopes that this new field will help them deal with personal and organizational realities that are both chaotic and complex.

This relationship between business and science goes back many years. In the seventeenth century, Newton's work was eagerly seized upon by the entrepreneurs giving birth to the Industrial Revolution (see Dobbs and Jacob 1995). Now, three hundred years later, Newtonian thinking continues to inspire the majority of our beliefs about the design and structure of organizations, as well as our theories about how to change them. In the early years of this century, science was brought deliberately into the nascent field of management theory. Marrying science with the art and craft of leadership was a way to give more credibility to this young and uncertain field. (This courtship continues today in full force, I believe from the same motivation.)

The work of Frederick Taylor, Frank Gilbreth, and hosts of followers initiated the era of "scientific management." This was the start of a continuing quest to treat work and workers as an engineering problem. Enormous focus went into creating time–motion studies and breaking work into discrete tasks that could be done by the most untrained of workers. I still find this early literature frightening to read. Designers were so focused on engineering efficient solutions that they completely discounted the human beings who were doing the work. They didn't just ignore them, as has been done more recently with contemporary reengineering efforts. They disdained them—their task was to design work that would not be disrupted by the expected stupidity of workers.

Though we in management may have left behind some of these beliefs and the rigid, fragmented structures that those beliefs engendered, we have not in any way abandoned science as the source of our credibility. Planning, measurement, motivation theory, organizational design and change—each of these and more bears the recognizable influence of science. Sometimes I see this traditional dependency most clearly when I listen to organizational theorists report on their research. I keep having the experience of going to professional conferences and listening to research reports that are rooted in seventeenth-century science. I am struck by how much we in the social sciences want to be seen as scientists. William Bygrave, a physicist who became a student of organizations, dubbed this "physics envy" (1989, 16). We feel afraid that we might lose our credibility without our links to math and physics, and I think this is true. Society demands this scientific standard, even as it turns around and criticizes these studies as too abstract and removed from the "real world."

In one presentation, an organizational trainer presented a long formula that captured, he assured us, all the relevant variables an employee would use to decide on further education. In another, a woman was assigning numerical values to relationships in a human network. She then plugged these numbers into a complex formula to assess the overall strength of the network. To be fair to these colleagues, I need to admit that in my professional life I have had a deep aversion to formulaic descriptions of human behavior. But I sat there aghast. There were their long strings of variables—separate descriptors interacting in precise, linear ways—and here was my brain, filled with my readings about nonlinearity, about chaos, about fuzzy particles that come into being as temporary relationships in the universal web. I was struck suddenly by the joke of it all. We social scientists strain for respectability, using the methodologies and thought patterns of seventeenth-century science, while the scientists, traveling away from us faster than the speed of light, are moving into a universe that calls for entirely new ways of understanding. Just when social scientists seem to have gotten the old math down, the scientists have left, plunging ahead into the vast "porridge of being" that describes a new reality.

Also to be fair, there are increasing numbers of social scientists experimenting with the nonlinear methodologies of new science, and many of them courageously do this research in spite of strong opposition from more traditional colleagues. I believe it is essential that this direction of research and application be supported. The science of the seventeenth century cannot explain what we are challenged by in the twenty-first century.

And it seems important for the social sciences to embrace new science for other reasons as well. Science still is the dominant thought form of our society. As scientist Lewis Thomas says, "Science affects the way we think together." We cannot escape its authority or deny the images it plants deep in the public imagination. Science is the voice that people hear. Yet as a social scientist, I have found it helpful to realize that I am working inside a powerful paradox. Many of the concepts that I and my colleagues are curious to understand are concepts that traditional science won't go near, such puzzles as identity, spirit, meaning, purpose, consciousness. Some scientists have directly addressed one or more of these concepts in their research, and others have called for a new epistemology of science that includes these domains as legitimate areas of scientific investigation (see Harman and Sahtouris 1998, Merchant 1980). While I have no idea whether science will eventually embrace these new questions, I do know that the power that science wields in our society draws me toward it. I am compelled to understand the vital science of our times.

Among its many influences, we can learn from new science to be more *playful,* to develop a different relationship with discovery. Nobel Prize winner Sir Peter Medawar said that scientists build "explanatory structures, telling stories which are scrupulously tested to see if they are stories about real life" (in Judson 1987, 3). I like this idea of storytellers. It works well to describe all of us. We are great weavers of tales, listening intently around the campfire to see which stories best capture our imagination and the experience of our lives. If we can look at ourselves truthfully in the light of this fire and stop being so serious about getting things "right"—as if there were still an objective reality out there—we can engage in life differently, more playfully. Lewis Thomas explains that he could tell something important was going on in an experimental laboratory by the laughter. Surprised by what nature has revealed, things at first always look startlingly funny. "Whenever you can hear laughter," Thomas says, "and somebody saying, 'But that's *preposterous!*'—you can tell that things are going well and that something probably worth looking at has begun to happen in the lab" (in Judson 1987, 71).

Wouldn't we all welcome more playfulness in our lives? I would be excited to encounter people delighted by surprises instead of being scared to death of them. Were we to become truly good scientists of our leadership craft, we would seek out surprises, relishing the unpredictable when it finally decided to reveal itself. Surprise *is* the only route to discovery, a moment that pulsates with new learnings. The dance of this universe requires that we open ourselves to the unknown. Knowing the steps ahead of time is not important; being willing to engage with the music and move freely onto the dance floor is what's essential.

One of the principles that guides scientific inquiry is that at all levels, nature seems to resemble itself. For me, the parsimony of nature's laws gives further impetus to my desire to learn from science. If nature uses certain principles to create her infinite diversity and her well-organized systems, it is highly probable that those principles apply to human life and organizations as well. There is no reason to think we'd be the exception. Nature's predisposition toward self-similarity gives me confidence that she can provide genuine guidance for the dilemmas of our time. We can use what we are learning in biology and physics to help us discern which current management ideas and practices are worth further inquiry. Science can help us develop new questions and processes that have merit at a more universal level. I feel better able to distinguish real nourishment from fast-food guru advice because of my awareness of the world that science now describes. Although I have intimated throughout these chapters some of the concepts from new science that illuminate organizational life, I'd like to highlight a few of them again.

For several decades now, there has been a growing chorus of research and practice that sings the praises of participative management. In reaction to this chorus, there are many critiques that describe the problems and shortcomings of participation. How can we know whom to believe? Is participation a fad that, like so many others, we can wait out, knowing it will pass? Is it based on democratic principles and is therefore non-transferable to other cultures? Is it merely a more sophisticated way to manipulate workers? Or is something else going on?

For me, new science answered those questions definitively. I believe in my bones that the movement towards participation is rooted in our changing perceptions of the organizing principles of life. Everywhere in the new sciences, in living systems theory, quantum physics, chaos and complexity theory, we observe life's dependence on participation. All life participates in the creation of itself, insisting on the freedom to self-determine. All life participates actively with its environment in the process of co-adaptation and co-evolution. No sub-atomic particle exists independent of its participation with other particles. And even reality is evoked through acts of participation between us and what we choose to notice.

As scientists fill us with images of this participatory universe, and even write about democracy as congruent with their science (see Kauffmann 1995 or Prigogine 1998), I wonder how we can continue to support authoritarian approaches. Can we resist inviting people to participate? Can we survive as command and control leaders? Can we hope that participation goes away? Not until life changes its fundamental processes.

The participatory nature of reality has required scientists to focus their attention on relationships. No one can contemplate a system's view of life without becoming engrossed in relational dynamics. Nothing exists independent of its relationships, whether looking at subatomic particles or human affairs. Certainly, relationships are a growing theme in today's leadership thinking. For many years, the prevailing maxim of management stated: "Management is getting work done through others." The important thing was the work; the "others" were distractions that needed to be managed into conformity and predictability. But now most of us have had to acknowledge that we are human, with our own insistent needs and gifts.

We tried for many years to avoid the messiness and complexity of being human, and now that denial is coming back to haunt us. We keep failing to create the outcomes and changes we need in organizations because we continue to deny that "the human element" is anything but a "soft" and not-to-be-taken-seriously minor distraction. We barely manage to survive the seemingly endless procession of organizational change fads and new ideas, each of which promises to make organizations more effective. CEOs acknowledge that about three-fourths of these efforts have failed. This terrible record of failure is, in my estimation, due to approaches that are predominantly technical and mechanistic. New technology is purchased; new organization charts are drawn; new training classes are offered. But most basic human dynamics are completely ignored: our need to trust one another, our need for meaningful work, our desire to contribute and be thanked for that contribution, our need to participate in changes that affect us.

Beyond the fads that have swept through large organizations, think of all the contemporary leadership problems that are variations on the theme that we don't know how to work together. We struggle to help teams form quickly and work effectively. We struggle to learn how to work with the uniqueness that we call diversity. We are terrified of the emotions aroused by conflict, loss, love. In all of these struggles, it is *being human* that creates the problem. We have not yet learned how to be together. I believe we have been kept apart by three primary Western cultural beliefs: individualism, competition, and a mechanistic world view. Western culture, even as it continues to influence people everywhere, has not prepared us to work together in this new world of relationships. And we don't even know that we lack these skills. In a simple example of the difficulties created by this ignorance, many MBA graduates who've been in the field a few years report that they wish they had focused more on organizational behavior and people skills while in school.

After all these years of denying the fact that we are humans, vulnerable to the same dynamics that swirl in all life (plus some unique to our species), we are being called to encounter one another in the messiness and beauty that name us as alive.

Many writers have offered new images of effective leaders. Each of them is trying to create imagery for the new relationships that are required, the new sensitivities needed to honor and elicit worker contributions. Here is a *very* partial list of new metaphors to describe leaders: gardeners, midwives, stewards, servants, missionaries, facilitators, conveners. Although each takes a slightly different approach, they all name a new posture for leaders, a stance that relies on new relationships with their networks of employees, stakeholders, and communities. No one can hope to lead any organization by standing outside or ignoring the web of relationships through which all work is accomplished. Leaders are being called to step forward as helpmates, supported by our willingness to have them lead us. Is this a fad? Or is it the web of life insisting that leaders join in with appropriate humility?

Participation and relationships are only two of our present dilemmas. Here we sit in the Information Age, the Knowledge Age, the Meaning Age—whatever it's called, we all feel besieged by more information than any mind can handle. Is information anything more than a new and perplexing tool given to us by technological advances? What if physicist John Archibald Wheeler is right? What if information is the basic ingredient of the universe? This is not a universe of things, but a universe of the "no-thing" of information. And this information is organized by a second invisible element, meaning. If the universe organizes through these invisible forces, then we must contemplate new processes for working with them. Information and meaning-making do not obey the classical laws of physics that govern matter. As energetic forces, they move and act differently—they can travel with great speed anywhere in the universal web and appear suddenly as potent influences that surprise us. In the West, we didn't grow up learning about non-material forces. But this has become a critical curriculum. We must learn how to work with life in all its dimensions, seen and unseen.

While information may be immaterial, we are all suffering under its weight. Information overload is a major problem. We aren't struggling with this problem just because of technology, and we won't solve our information dilemmas just by using more sophisticated information-sorting techniques. Something much grander is being asked of us. We are moving irrevocably into a new relationship with the creative element of life. However long we may hope it isn't true, we will be forced to accept that information—freely generated, freely communicated, and freely interpreted—is our only hope for self-organized order in a world that no longer waits for us to respond. If we fail to recognize information's essential role in supporting self-organization, we will be unable to survive in this new world.

Information needs to be free, and the necessity for freedom is another prevailing message in much of new science. This world insists that we develop a different understanding of autonomy and self-determination, moving far from the command-and-control approaches of the past. To many managers, autonomy is just one small step away from anarchy. They hesitate to use it unless they can be assured it will be carefully controlled. As one manager wryly commented, "I believe in fully autonomous work, as long as it stops at the level below me." Yet everywhere in nature, the freedom to self-determine is essential. What's peculiar about this freedom is that it results not in anarchy, but in global systems that support all members of the system. Individuals and local groups are free to do what makes sense to them. These local units respond, adapt, change. Another manager put it succinctly: "People need to be free to do what has to get done."

What emerges from this freedom is a globally stable system. Rather than building a rigid organization piece by stable piece, nature keeps things freely moving at all levels. These movements emerge into something new—an integrated system that can resist most demands for change at the global level because there is so much internal motion.

The motion of these systems is kept in harmony by life's great cohering process, that of self-reference. While new in science, self-reference has been an enduring concept in human thought. In Greek times, the Delphic Oracle greeted supplicants with this principle engraved in marble: "Know Thyself." And Shakespeare counseled, "This above all, to thine own self be true." So contemporary science is merely bringing to light a wisdom that has been with us for millennia. We see the world through who we are. All living beings create themselves and then use that "self" to filter new information and co-create their worlds. We refer to this self to determine what's important for us to notice. Through the self, we bring form and meaning to the infinite cacophony of data that always surrounds us.

Yet it is very important to note that in all life, the self is not a selfish individual. "Self" includes awareness of those others it must relate to as part of its system. Even among simple cells, there is an unerring recognition that they are in a *system*; there is a profound relationship between individual activity and the whole.

In a living system, self-reference is the source of growth, of increasing vitality. But for machines, it doesn't work that way. *Star Trek* popularized an effective method for destroying computers; you program them with a self-referential statement, such as "Prove that your prime directive is not your prime directive" (Briggs and Peat 1989, 67). As the logic turns back on itself in unending iterations, a machine will blow its circuits. Zen masters employ the same technique with koans, but they know that humans are not machines and that we can be challenged to new levels of insight by self-referential exercises. As we abandon the machine imagery of the past, self-reference calls to me as the richest and most enticing teacher for how to be together in ways that support life, not destruction.

Self-reference conjures up such different possibilities for how to be together. It explains how life creates order without control, and stable identities that are open to change. It describes systems of relationships where both interdependence and individual autonomy are necessary conditions. It promises that as individuals together reference a chosen, shared identity, a coherent system can emerge. It illuminates the necessity for meaning-making in a world that often feels meaningless.

But before we can embrace this fundamental life process, we need to explore a more elemental issue. We need to determine whether we, each of us, believe that this is an orderly universe. For me, it is not only the science I have read that gives me assurance that I live in an orderly world, even when it refuses to organize in ways of my choosing. I have spent years trying to see differently, to look for order and the processes by which newness comes into form. Being out in the world with new eyes and a willingness to be taught, I have found that nature and people provide more hopeful examples of self-organization than I can possibly comprehend.

For me, there is no choice but to continue on the path new science has helped mark. Like all journeys, this one moves through both the dark and the light, the terrors of the unknown and the joys of deep recognition. Some shapes and landmarks are already clear. Others wait to be discovered. No one can say where the journey is leading. But the relationship promises to be fruitful, and I can feel the explorer's blood rising in me. I am glad to feel in awe again.

Chapter Ten

The Real World

P eople often comment that the new leadership I propose couldn't possibly work in "the real world." I assume they are referring to their organization or government, a mechanistic world managed by bureaucracy, governed by policies and laws, filled with people who do what they're told, who surrender their freedom to leaders and sit passively waiting for instructions. This "real world" craves efficiency and obedience. It relies on standard operating procedures for every situation, even when chaos erupts and things spin out of control.

This is not the real world. This world is a manmade, dangerous fiction that destroys our capacity to deal well with what's really going on. The real world, not this fake one, demands that we learn to cope with chaos, that we understand what motivates humans, and that we adopt strategies and behaviors that lead to order, not more chaos.

In this historic moment, we live caught between a worldview that no longer works and a new one that seems too bizarre to contemplate. To expose this, I want to apply the lens of new science to two of society's most compelling, real world challenges: How well we deal with natural and manmade disasters and how well we respond to global terror networks. Using this high-resolution lens, we can see many dynamics that are crucial to understand, yet were obscured from view by our old sight.

Here is the real world described by new science. It is a world of interconnected networks, where slight disturbances in one part of the system may create major impacts far from where they originate. In this highly sensitive system, the most minute actions can blow up into massive disruptions and chaos. But it is also a world that seeks order. When chaos erupts, it not only disintegrates the current structure, it also creates the conditions for new order to emerge. Change always involves a dark night when everything falls apart. Yet if this period of dissolution is used to create new meaning, then chaos ends and new order emerges.

This is a world that knows how to organize itself without command, control, or charisma. Everywhere, life self-organizes as networks of relationships. When individuals discover a common interest or passion, they organize themselves and figure out how to make things happen. Self-organizing evokes creativity and results, creating strong, adaptive systems. Surprising new strengths and capacities emerge.

In this world, the "basic building blocks" of life are relationships, not individuals. Nothing exists on its own or has a final, fixed identity. We are all "bundles of potential." Relationships evoke these potentials. We change as we meet different people or are in different circumstances.

And strangest of all, scientists cannot find any independent reality that exists without our observations. We create reality through our acts of observation. What we perceive becomes true for us and this version of reality becomes the lens through which we interpret events. This is why we can experience the same event or look at the same information and have very different descriptions of it.

This real world stands in stark and absolute contrast to the world invented by Western thought. We believe that people, organizations, and the world are machines, and we organize massive systems to run like clockwork in a steady-state world. The leader's job is to create stability and control, because without human intervention, there is no hope for order. Without strong leadership, everything falls apart. It is assumed that most people are dull, not creative, that people need to be bossed around, that new skills develop only through training. People are motivated using fear and rewards; internal motivators such as compassion and generosity are discounted. These beliefs have created a world filled with disengaged workers who behave like robots, struggling in organizations that become more chaotic and ungovernable over time.

And most importantly, as we cling ever more desperately to these false beliefs, we destroy our ability to respond to the major challenges of these times.

Leadership in Disasters: Learning from Katrina

The world has experienced so many disasters and human tragedies in the past several years that some worry about "compassion fatigue." I don't believe that our compassion is finite and in danger of being exhausted. The source of our fatigue is that we don't have the organizational structures or the leadership that can respond quickly and well to these emergencies. We want to help, but our organizations fail to deliver our compassion to those most in need. This is both frustrating and exhausting because, as humans, we are spontaneously generous and want to be of service.

Following any disaster, we see the best of human nature and the worst of bureaucracy. Headlines convey our frustration: "Poor Nations Say Much Charity Fails to Reach Victims," "System Failure: An Investigation into What Went so Wrong in New Orleans," "Red Cross Under Investigation," "Congress Probe Examines What Went Wrong."

Other headlines speak to the valiancy of individuals and unofficial relief efforts : "Real-life Heroes," "Organized Churches Are Not an Oxymoron," "No Red Cross, No Salvation Army or Federal Funds . . . Just Friends."

Time Magazine relayed this story in September 2005 just weeks after hurricane Katrina struck the Gulf Coast. It illustrates the conflict between willing volunteers and government bureaucracy.

> As flames blazed 400 miles away in New Orleans on Labor Day, about 600 firefighters from across the nation sat in an Atlanta hotel listening to a FEMA lecture on equal opportunity, sexual harassment, and customer service. "Your job is going to be community relations," a FEMA official told them . . . "You'll be passing out FEMA pamphlets and our phone number."
>
> The room, filled with many fire fighters who, at FEMA's request, had arrived equipped with rescue gear, erupted in anger. "This is ridiculous," one yelled back. "Our fire departments and mayors sent us down here to save people, and you've got us doing this?" The FEMA official climbed atop a chair . . . and tried to restore order. "You are now employees of FEMA, and you will follow orders and do what you're told," he said, sounding more like the leader of an invading army than a rescue squad. . . .
>
> [The firefighters] got tired of hanging around their hotel and returned home (*Time*, 2005, 39).

Although this story is appalling, it happens all the time in disasters. The first response of people is to do anything they possibly can to help, rescue, and save other people. They gather resources, invent solutions on the spot, and work tirelessly for days on end. They don't think about risk or reward—these are spontaneous outpourings of compassion focused creatively and purposefully. A group of mid-level managers for Southwest Bell described how they felt responding to the Oklahoma City bombing: "There was no risk. It was already a disaster."

Yet these self-organized efforts are often hindered by officials who refuse their offers, cite regulations, or insist that protocols and procedures be followed. This is not a criticism of individual officials—they are imprisoned by their role and can't act independently. As *Time* Magazine described what happened with Hurricane Katrina: " . . . at every level of government there was uncertainty about who was in charge at crucial moments. Leaders were afraid to actually lead, reluctant to cost businesses money, break jurisdictional rules, or spawn lawsuits. They were afraid, in other words, of ending up in an article just like this one" (2005, 36).

Concerns about who had legal and decision-making authority created many nightmares. Official requests for aid were given to the wrong person or to someone who didn't understand it and denied the request. If requests were made to the right desk but not worded correctly, they were ignored or denied. The Louisiana governor requested Federal help from the President. When asked what she needed, she replied: "Give me all you got." That plea was not deemed sufficient for the Federal government to step in, and days passed before Federal and state officials worked out who had jurisdictional authority (*Time*, 2005).

As people argued about their roles and authority, no one saw the pattern of destruction and chaos that was unfolding. Officials responded only to the disconnected bits of information that related directly to their offices. No one seemed to understand the information they were getting, or notice that they were only seeing a small portion of what was happening. There were many instances when images of terrified, suffering people filled TV screens while on another channel, government officials denied there were any serious problems. In some cases, their inability to comprehend what was happening was due to inexperience (from job turnover). In other cases, the problem was a new chain of command, with managers focused on terrorism in the Department of Homeland Security now responsible for FEMA, yet had no understanding of natural disasters.

Even before Katrina hit, key decision-makers at all levels of government displayed a curious blindness. Years of simulations and analyses had created clear descriptions of the damage that would result from a category 3 or 4 hurricane. The destruction of New Orleans was one of the top three potential catastrophes listed by FEMA for many years. How is it possible that officials were blindsided and failed to prepare adequately for this eventuality? And why were they so slow to respond, even as the National Weather Service mapped Katrina's approach with unerring accuracy? It was as if government officials at all levels could not comprehend the reality of what was about to happen. Either they discounted the information, failed to interpret it correctly, or duped themselves into believing "it can't happen here." This is a familiar yet troubling example of paradigm blindness, where people are unable to see information that threatens and disconfirms their worldview. No matter how much data is in front of them, their lens filters it out or distorts it to mean something else. And in some cases, people literally do not see the information, even if it's right in front of them (see Kuhn 1969).

In the days after Hurricane Katrina, this blindness was coupled with bureaucratic conditioning and cumbersome chains of command. Missteps, misperceptions, and inaction cascaded through organizations, creating only more chaos. An already devastating set of circumstances turned even more tragic because of the failure of leaders to perceive accurately what was going on and to risk taking actions that went beyond the bonds of bureaucracy.

However, all along the Gulf Coast, people self-organized with neighbors and strangers to help and rescue people. The efforts of amateur ham radio operators created an immediate and effective communications network that saved many lives. In one case, a desperate family in New Orleans could not get any response from their local 911 number. They did, however, reach a relative a thousand miles away. He called *his* local 911, who contacted a New Orleans ham operator, who relayed the information to local people, who then rescued the family (Sky, 2006).

Unlike official agencies, many of these operators prepared themselves ahead of time. They established themselves in safe, dry places before the storm struck. Acting independently, each with their own generator and transmitter, they wove a powerful network of communications. Their independence made them extremely resilient. If one person could no longer transmit, another picked up quickly. "Each one is a mobile, independent unit working in cooperation for a common goal" (Sky, 83). They acted freely, but from a clearly shared intent. These are the conditions that make it possible to bring order out of chaos.

 Senior leaders find it difficult to act this spontaneously or independently. Any independent response is constrained by the need to maintain the power and policies of the organization. Paralyzed by formal operating procedures, it takes courage to forego these controls and do what you think might help. The Southwest Bell employees in Oklahoma City leapt into action immediately after the bombing of the Federal Building, in large part because their leaders were out of town. When the leaders returned, their staff told them: "It's good you weren't here. We could just take action." Although this is never what a leader wants to hear, these leaders were wise enough to know this was true and that their absence had created value.

In Hurricane Katrina, the chain of command and the observance of protocol created even more disasters:

> While people were dying in New Orleans, the U.S.S. *Bataan* steamed offshore, its six operating rooms, beds for 600 patients and most of its 1,200 sailors idle. Foreign nations . . . readied rescue supplies, then were told to stand by for days until FEMA could figure out what to do with them. Florida airboaters had an armada ready for rescue work but complained that FEMA wouldn't let them into New Orleans. Brown defended his agency's measured steps, saying aid "has to be coordinated in such a way that it's used most effectively" (*Time*, 39–40).

Leaders who respond quickly ignore standard operating procedures. In the state of West Virginia, the governor didn't wait to be asked but immediately mobilized six C-130 cargo planes from the National Guard to go and pick up people needing evacuation. The planes were sent filled with supplies and were expected to return filled with people. The governor was there to welcome them when they arrived, but only three planes came back with people. FEMA had refused to let more people board the planes. About 400 evacuees benefited from this quickly mobilized relief effort. Although economically poor, West Virginia offered more assistance than their affluent neighboring states, all because they rallied around the governor's call to help brothers and sisters whom they had never met.

In contrast to the terrible failures of government, communities, individuals, and small groups responded immediately to Katrina. One commentator describes these responses as "acts of love in times of danger" (*The Nation*, 2005,13). The community of Ville Platte exemplified the generous self-organizing capacity that always appears in disasters (*The Nation*, 13-18). They organized their "home-made rescue and relief efforts" around the slogan "If not us, then who?" A community of 11,000 people, with an average yearly income of only $5300 for the majority of its residents, was able to serve 5000 displaced and traumatized victims of Katrina, inviting them to share their homes and community not as refugees or evacuees, but as "company." Those with boats went to New Orleans to join "The Cajun Navy." They rescued people from rooftops, picked up the dead, transported the injured to trauma centers. They saw people from other communities doing the same thing. FEMA wasn't around, "That was it. Just us volunteers."

Ville Platte helped thousands of "company" without any Federal or Red Cross aid (they did try to reach the Red Cross, but gave up after thirteen days of calling with no response.) Their success cannot be explained by the old mechanical paradigm, but is easily understood by the dynamics described in new science. We live in a world of relationships, where each event or person evokes new capacities. We live in a world where order emerges out of chaos if people are free to make their own decisions based on shared meaning and values. We live in a world where effective response doesn't require top-down leadership or an organization plan drawn up ahead of time. People self-organize in order to accomplish something that matters to them. As one community member said: "All of us know how to spontaneously cooperate. My God, we're always organizing christenings or family gatherings. So why do we need a lot of formal leadership?"

In a disaster, where quick response is demanded, formal organizations are incapacitated by the very means they normally use to get things done— chains of command, designated leaders, policies, procedures, plans, regulations, and laws. We *can* rely on human compassion, but we need to develop the means for official agencies to support, work with, and not resist the self-organizing capacity of people that always emerges in a disaster. Leaders need to have the freedom to make intelligent decisions based on their comprehension of the situation, not their understanding of policies and procedures. The formal leader's job is to ensure that the resources they control get to local groups as fast as possible. Leaders need to trust that people will invent their own solutions, that they'll make good use of the resources they can provide. And leaders need to expect and value the unique and inventive responses created in each community, rather than enforcing compliance to one-size-fits-all.

These radically different behaviors require that we free official leaders to act wisely, and that they trust people to self-organize effective responses. How much more sad history do we have to repeat before we understand this? Let us hope we learn from Katrina that the only way to restore order out of chaos is to rely on people's intelligence, love, and capacity to self-organize, to accomplish what they care about.

We also need to entrust local people with official resources of money and materials for the rebuilding. When rebuilding is left to governments, outside contractors, and large nonprofit organizations, progress gets mired down in regulations, time drags on, people's needs aren't served, and no one from the local community is satisfied with the results. Supporting initiatives where local people do the work sustains local cultures, recreates community cohesion, and is accomplished at amazing speed. The clean-up of the World Trade Center's Ground Zero was accomplished in record time, with no traditional New York and contractor politics; people worked overtime and risked their health to remove the debris of their shared tragedy.

In the 1990s, almost two billion people were affected by disasters, 90% of them in the most impoverished nations. We will not succeed in responding effectively, and in ways that satisfy our compassion, until we change how we organize relief efforts. The basic shift needs to be from control to order, from a reliance on formal authority and procedures to a reliance on the self-organizing capacities of local people, agency staffers, and those who volunteer to help. Some of the more progressive thinking on disaster relief focuses on how to mobilize and develop local people by engaging them in the work of rescue and rebuilding. If local people are engaged, they "move from object to subject, victim to actor, to the possibility of being (Smillie, 2001)."

This capacity to create solutions without traditional hierarchies or formal leadership is found in communities everywhere, not just those facing disasters. At The Berkana Institute, (which I co-founded in 1992) we work with the assumption that "the leaders we need are already here." We have discovered that even in the most economically poor communities in the world there is an abundance of leaders. These leaders work to strengthen their community's ability to be self-reliant by working with the wisdom and wealth already present in its people, traditions and environment (see pages 196–197). A 2002 Ford Foundation report on leadership notes the same thing. "There is a sense among some in our country today that we are lacking inspirational leaders. . . . Yet a closer look reveals that all over the nation groups of concerned citizens are working together, often at the local level, to solve tough social problems. These are the new leaders in America today (Louv, 2002)."

We need to consider carefully what we are learning about leadership in these disaster-laden times. I hope we learn that we *can* rely on human caring, creativity, and compassion. We *can* rely on people as 'bundles of potential' figuring out solutions, learning quickly, and surprising ourselves with new capacities. We *can* rely on people to self-organize quickly to achieve results important to them. Together, people act creatively, take risks, invent, console, inspire, and produce. This is how life works. We can learn this from new science, or we can learn it from what happens everyday somewhere in the real world.

Leadership of Networks: Learning from Terrorist Groups

How is it possible that a few thousand enraged people can threaten the stability of the world? How is it possible that the most powerful governments on earth find themselves locked in a costly and fearsome struggle, diverting large amounts of resources and attention to suppress the actions of a small group of fanatics? It's hard to acknowledge the power and success of global terror networks, but they are among the most effective and powerful organizations in the world today, capable of changing the course of history. They do this without formal power, advanced technology, huge budgets, or large numbers of followers.

What are the criteria we use to judge effective leaders? They include the abilities to communicate a powerful vision, to motivate people to work hard for them, to achieve results, exceed plans, and implement change. We want their leadership to result in a resilient organization able to survive disruptions and crises, one that grows in capacity, that doesn't lose its way even after the leader retires. If we apply these criteria to the leaders of terrorist networks, they come out with high marks. It's difficult to acknowledge them as our teachers, but we have much to learn from them about innovation, motivation, resiliency, and the working of networks.

New science explains the behavior of networks in great detail because this is the only form of organization used by the planet. With the lens of science, we can peer into these terrorist organizations and explore the methods of their success. We can also see how to respond in ways that ensure we stop contributing to their success.

At present, we are dangerously blind to their strength because we use the wrong lens to evaluate their capacity. We use factors that apply to our world but not to theirs; to the behavior of hierarchical organizations, not to networks. Failing to use the right lens, we think we are winning the war on terror. We ask whether Osama bin Laden is still a threat, whether Al-Qaeda is losing its strength, by evaluating his ability to give orders or to communicate using advanced technology. We assume that bin Laden is a weaker leader now that he is on the run and hiding in caves. We assume that if we prevent communication, terrorists won't be given orders and therefore won't launch attacks. We assume that if we kill the top leaders, if we decapitate their organization, that young terrorists will slink away from this anarchic, leaderless group.

U.S. military commanders frequently acknowledge they are fighting a new kind of enemy. They describe this enemy as one who learns, changes, adapts. As soon as U.S. soldiers figure out insurgents' strategy, it is changed. Think about the vast resources nations spent on defending themselves against the *last* terrorist attack, even though experience teaches that terrorists never repeat themselves.

The Army's long-term strategy is to develop a fighting force that is as adaptive, nimble, and smart as the insurgents. (The ten-year plan is to develop many more Special Forces.) The military has studied the behavior of networks and the emergence of "netwars" for many years. Before 9/11, they warned of the proliferation of networks; not only transnational terrorist groups, but also black market sales of WMDs, drug and crime syndicates, fundamentalist and ethno-nationalist movements, immigration smugglers, urban gangs, back-country militias and militant single-issue groups (Arquilla 2001, 6). As networks, these groups operate in small, dispersed units that can deploy nimbly—anywhere, anytime. They know how to penetrate and disrupt, as well as elude and evade. Many groups are leaderless (Arquilla, ix). They also attack by "swarming," suddenly appearing from multiple directions, coalescing quickly and secretly, then disintegrating as quickly as they appeared (Arquilla,12, also Rheingold).

Although these groups appear leaderless, they in fact are well-led by their passion, rage, and conviction. They share an ideal or purpose that gives them a group identity and which compels them to act. They are geographically separate, but "all of one mind" (Arquilla, 9). They act free of constraints, encouraged to do "what they think is best" to further the cause. This combination of shared meaning with freedom to determine one's actions is how system's grow to be more effective and well-ordered. The science thus predicts why terrorist networks become more effective over time. If individuals are free to invent their own ways to demonstrate support of their cause, they will invent ever more destructive actions, competing with one another for the most spectacular attack.

People who are deeply connected to a cause don't need directives, rewards, or leaders to tell them what to do. Inflamed, passionate, and working with like-minded others, they create increasingly extreme means to support their cause. Describing Al Qaeda's success, network analyst Albert-László Barabási notes: "Bin Laden and his lieutenants did not invent terrorist networks. They only rode the rage of Islamic militants, exploiting the laws of self-organization along their journey (2002, 224)." An insurgency is not "as is often depicted, a coherent organization whose members dutifully carry out orders from above, but a far-flung collection of smaller groups that often act on their own or come together for a single attack (*The New York Times* 12/2/05)." In this way, movements that begin as reasonable often migrate to more extremist measures, propelled there by their members' zealousness. And with passions inflamed, growth is assured. The dramatic acts of one small group inspire many copycat actions in places far distant.

Over time, a network is fueled more by passion than by information. Networks begin with the circulation of information. This is how members find each other, learn from each other, and develop strategies and actions. Most attempts for disrupting network activities focus on how to interfere with their communications. But once the network has momentum, passion and individual creativity propel it forward. Communication is still essential for large coordinated attacks, but the proliferation of small, disconnected, lethal attacks does not require information. It only requires passionate commitment and a willingness to martyr oneself. Therefore, as the anger of network members grows in intensity, information plays a lesser role, and personal innovation takes over. When we succeed in disrupting network communications, we also incite more local rage. Individuals may not be able to communicate with each other but, in their isolation, they become more creative in designing their own deadly attacks. So we can never measure adequately our success in disrupting a network by measuring only how well we are disrupting their communications.

The essential structure of any network is horizontal, not hierarchical, and ad hoc, not unified. This broad dispersal makes it difficult to suppress any rebel group. "Attack any single part of it, and the rest carry on largely untouched. It cannot be decapitated because the insurgency, for the most part, has no head (*The New York Times,* 12/2/05)." What appears as atomized and fragmented is, in fact, far more lethal than an organized military force. Bruce Hoffman, a Rand

Corporation expert on terrorism states: "There is no center of gravity, no leadership, no hierarchy; they are more a constellation than an organization. . . . They have adopted a structure that ensures their longevity" (*The New York Times,* 12/2/05)."

These descriptions and dynamics do not surprise anyone familiar with new science and its observations of networks. Networks possess amazing resiliency. They are filled with redundant nodes, so that one picks up if another goes down (as did the ham radio operators in New Orleans.) And human networks always organize around shared meaning. Individuals respond to the same issue or cause and join together to advance that cause. For humans, meaning is a "strange attractor"—a coherent force that holds seemingly random behaviors within a boundary. What emerges is coordinated behaviors without control, and leaderless organizations that are far more effective in accomplishing their goals.

When we think of organizations as machines, we are blind to the power of self-organized networks. We keep looking for the leader. We assess an insurgency by whether its leader is visible, available, and able to communicate easily with the forces. This is a profound and dangerous misperception of the leader's role. In early 2006, I listened to interviews with U.S. analysts trying to assess whether bin Laden was still a threat. They were looking at traditional organizational attributes: visibility, technology, chain of command, ability to issue orders, communication channels. Against those criteria, it seemed that bin Laden's power had been severely reduced. But one network expert said: "It's the idea, not the organization. . . bin Laden is a person of influence" (National Public Radio 1/25/06, Morning Edition). And Barabási warns that: "Because of its distributed self-organized topology, Al Qaeda is so scattered and self-sustaining that even the elimination of Osama bin Laden and his closest deputies might not eradicate the threat they created. It is a web without a true spider" (2002, 223).

The science of how networks emerge out of chaos, organize around shared meaning and grow more effective provides new and more accurate measures for assessing the strength of Al Qaeda and other insurgencies. These measures focus not on size, structure or chain of command, but on meaning and emotions. They are startlingly different to the traditional ones we use.

1. Instead of counting the number of insurgents, how can we assess their passion and rage? A rise in attacks and demonstrations indicates increasing rage.

2. Is there a predictable pattern to attacks? Or are they becoming more varied? Greater variety of attacks indicates local initiative. This indicates increased dedication to the cause and less reliance on a central authority.

3. Where are attacks occurring? More attacks in surprising places is evidence of the network's strength, that it is growing.

4. What is the impact of our actions in fueling the passion of network members? Is what we're doing fanning the flames or working to pacify the situation?

5. To determine the leader's influence, look at the popularity of his ideas and interpretations. Do people accept his interpretations without question or do they debate them? How does the leader's appearance (in any form) affect the behavior of his followers? Is there any correspondence between the number of attacks and these announcements? Or do attacks continue to escalate independent of his presence? If attacks increase without his visibility, this indicates the network's momentum, "a web without a spider."

6. To determine the network's resiliency, what happens when a node or cell is destroyed? Have the number of attacks decreased or just shifted to a new location?

These and other measures would lead to a very different assessment of who is winning the war on terror. If networks grow from passion, if Al Qaeda "rides the rage" of angry Islamic militants, then the best strategy for immobilizing terrorist networks is not to kill their leaders, but to defuse the sources of their anger, and not to incite them further. Many analysts arrive at a similar conclusion—we can only win the war on terror by eliminating the causes of rage. As long as our actions provoke their anger, we can expect more terrorists, more extreme attacks, and the continuing destablilization of the world by a small group of people. Barabisi states: "If we ever want to win the war, our only hope is to tackle the underlying social, economic, and political roots that fuel the network's growth. We must help eliminate the need and desire . . . to form links to terrorist organizations by offering them a chance to belong to more constructive and meaningful webs." We might win small and discrete battles, we might break up different cell groups, but if we do nothing to eliminate their rage, people will continue to form these deadly networks and "the netwar will never end" (224).

Similar clarity pervades the work of military strategist and advisor Thomas Barnett, who links economic progress to national security. Barnett notes that one-third of humanity lives outside the global economy in "the Gap." Their economic poverty has serious consequences because, since the end of the Cold War, " all the wars and civil wars and genocide have occurred within the Gap." To achieve true security, we must ensure that these populations benefit from economic advantages, thus "eradicating the disconnectedness that defines danger in the world today" (2005, xii).

This is the real world that we resist seeing at our own imminent peril. If we continue to seek to control it by exerting ever more pressure on those who hate us, those who feel disconnected, those who are impoverished, we only create a future of increasing disorder and terror. But to see a new way out of this terrifying future, we must learn to understand and see the world differently. Einstein's wonderful counsel that no problem is ever solved by the same thinking that created it defines what we must do. We must understand the behavior of networks in this densely interconnected world. We must understand human motivation and our astonishing capacity to self-organize when we care about something. We must understand that we lose capacity and in fact create more chaos when we insist on hierarchy, roles, and command and control leadership.

There is no more time to think about whether we need to make this shift. We can't afford to continue wandering blindly in the real world, oblivious to what's going on. But if we can become curious and willing students of life's dynamics, I know we will discover surprising new capacities and insights. Whenever we humans see clearly and understand the true dimensions of any problem, we become brave and intelligent actors in the world. It is time to open our eyes, change our lens, and step forward into actions that will restore sanity and possibility to the real world.

This is the setting out.

The leaving of everything behind.

Leaving the social milieu. The preconceptions. The definitions. The
*   language. The narrowed field of vision. The expectations.*

No longer expecting relationships, memories, words, or letters to mean
*   what they used to mean. To be, in a word: Open.*
                                        —*Rabbi Lawrence Kushner*

Journeying to a New World

Across the valley, the last colors of this day warm the horizon. Two dimensions move across the land, removing all contours, smoothing purple mountains flat against a rose-radiant sky. Whenever natural forces of destruction are active anywhere in Asia, the skies of Utah light up. At every twilight, visiting dust shimmers red in the air, intensifying the colors of an always intense sky. I sit bathed in strange light, anchored by dark magenta mountains.

I move differently in the world these days since traveling in the realms of new science. The world has become a strange and puzzling place that keeps insisting I give up what I thought I knew. But I find life much more interesting now, living with not knowing, trying to stay curious rather than certain. In the process of writing this book, of playing with its ideas for a number of years, and then rewriting it based on what I've seen, a few things about the journey stand out.

I was in this work a few years before I was able to identify its real nature. I realized that I and others weren't asking people simply to adopt some new approaches to leadership, or to think about organizations in a few new ways. What we were really asking, and what was also being asked of us, was that we change our thinking at the most fundamental level, that of our world view. The dominant world view of Western culture—the world as machine—doesn't help us to live well in this world any longer. We have to see the world differently if we are to live in it more harmoniously.

Once I understood the nature of the work, it helped me relax and be more generous. I learned that people get frightened if asked to change their world view. And why wouldn't they? Of course people will get defensive; of course they might be intrigued by a new idea, but then turn away in fear. They are smart enough to realize how much they would have to change if they accepted that idea. I no longer worry that if I could just find the right words or techniques, I could instantly convince people. I no longer expect a new world view to be embraced quickly; I don't know if I'll see it take root in my lifetime. I also know that people are being influenced from sources far beyond anyone's control. I know many people who've been changed by events in their lives, not by words they read in a book.

These people have been changed by life's great creative force, chaos. One of the gifts offered by this new world view is a clearer description of life's cyclical nature. The mechanistic world view promised us lives of continual progress. Since we were in control and engineering it all, we could pull ourselves straight uphill, scarcely faltering. But life doesn't work that way, and this new world view confirms what most of us knew—no rebirth is possible without moving through a dark passage. Dark times are normal to life; there's nothing wrong with us when we periodically plunge into the abyss.

Over the past years, nudged by the science, I have come to know personally that the journey to newness is filled with the black potholes of chaos. The science has restrained me from trying to negotiate my way out of dark times with a quick fix. But even though I know the role of chaos, I still don't like it. It's terrifying when the world I so carefully held together dissolves. I don't like feeling lost and emptied of meaning. I would prefer an easier path to transformation. But even as I experience their demands as unreasonable, I know I am in partnership with great creative forces. I know that chaos is a necessary place for me to dwell occasionally. So I have learned to sit with these dark moments—confused, overwhelmed, only faintly trusting that new insight will appear. I know that this is my only route to new ways of being.

The more I contemplate these times, when we truly are giving birth to a new world view, the more I realize that our culture is presently journeying through chaos. The old ways are dissolving, and the new has not yet shown itself. If this is true, then we must engage with one another differently, as explorers and discoverers. I believe it will make the passage more fruitful if we can learn how to honor each other in these roles. We can realize that no single person or school of thought has the answer, because what's required is far beyond isolated answers. We can realize that we must inquire together to find the new. We can turn to one another as our best hope for inventing and discovering the worlds we are seeking.

In the past, exploration was easier. We could act as patrons and pay somebody to do it for us. *They* would set sail and bring back to us the answers and riches we coveted. We still want it to work this way; we still look to take what others have discovered and adopt it as our own. But we have all learned from experience that solutions don't transfer. These failures have been explained by quantum physics. In a quantum world, everything depends on context, on the unique relationships available in the moment. Since relationships are different from place to place and moment to moment, why would we expect that solutions developed in one context would work the same in another?

So we can no longer act as patrons, waiting expectantly for the right solution. We are each required to go down to the dock and begin our individual journeys. The seas need to be crowded with explorers, each of us looking for our answers. We *do* need to be sharing what we find, but not as models. From each other, we need to learn what's possible. Another's success encourages us to continue our own search for treasure.

This need to discover for ourselves is unnerving. I keep hoping I'm wrong and that someone, somewhere, really *does* have the answer. But I know we don't inhabit that universe any longer. In this new world, you and I have to make it up as we go along, not because we lack expertise or planning skills, but because that is the nature of reality. Reality changes shape and meaning as we're in it. It is constantly new. We are required to be there, as active participants. It can't happen without us, and nobody can do it for us.

If we take seriously the role of explorer and inventor, we will realize that we can't do this alone. It's scary work, trying to find a new world, hoping we won't die in the process. We live in a time of chaos, as rich in the potential for disaster as for new possibilities. How will we navigate these times?

The answer is, together. We need each other differently now. We cannot hide behind our boundaries, or hold onto the belief that we can survive alone. We need each other to test out ideas, to share what we're learning, to help us see in new ways, to listen to our stories. We need each other to forgive us when we fail, to trust us with their dreams, to offer their hope when we've lost our own.

I crave companions, not competitors. I want people to sail *with* me through this puzzling and frightening world. I expect to fail at moments on this journey, to get lost—how could I not? And I expect that you too will fail. Even our voyage is cyclical—we can't help but move from old to new to old. We will vacillate, one day doing something bold and different, excited over our progress, the next day, back to old behaviors, confused about how to proceed. We need to expect that we will wander off course and not make straight progress to our destination. To stay the course, we need patience, compassion, and forgiveness. We should require this of one another. It will help us be bolder explorers; it might keep us from going mad.

This is a strange world, and it promises only to get stranger. Niels Bohr, who engaged with Heisenberg in those long, nighttime conversations that ended in despair, once said that great ideas, when they appear, seem muddled and strange. They are only half-understood by their discoverer and remain a mystery to everyone else. But if an idea does not appear bizarre, he counseled, there is no hope for it (Wilber 1985, 20). So we must live with the strange and the bizarre, directed to unseen lands by faint glimmers of hope. Every moment of this journey requires that we be comfortable with uncertainty and appreciative of chaos' role. Every moment requires that we stay together. After all is said and done, we have the gift of each other. We have each other's curiosity, wisdom, and courage. And we have Life, whose great ordering powers, if we choose to work with them, will make us even more curious, wise, and courageous.

Illustration Credits

Acknowledgment is made to the following sources for permissions to use:
35, bubble chamber, courtesy of the Lawrence Berkeley Laboratory, University of California; **117**, reprinted by permission of the William Morris Agency, Inc. on behalf of the author © 1987 by James Gleick; **124**, illustration by Lynn Farrar; **127**, The Chaos Game from *The Fractal Explorer,* Dynamic Press, Santa Cruz, CA; illustration by Linda Garcia 1991; **134**, illustration by Lynn Farrar.

The color plates:
Plate 1, Three-winged Bird: A Chaotic Strange Attractor, Mario Markus and Benno Hess, Max-Planck-Institut, Dortmund, Germany; **Plates 2 and 3**, Julia Set fractal © 1992–Lifesmith Classic Fractals, Northridge, CA 91324, USA; **Plate 4** (top) Clouds, photograph by R. Blair; **Plate 4** (bottom) Grand Canyon from Toroweap, courtesy of Brigham Young University, photograph by Frank Jensen; **Plate 5** (top) Computer generated ferns © 1992–Lifesmith Classic Fractals, Northridge, CA 91324, USA; **Plate 5**, (bottom) courtesy of Brigham Young University; **Plate 6**, Belousov-Zhabotinsky Reaction, courtesy of Brigham Young University; **Plate 7** (top), Hurricane Edna, photographed from the space shuttle, courtesy National Weather Service; **Plate 7** (bottom), Copper double spiral ornament photograph by Kenneth Garrett/NGS Image Collection, courtesy of the Swiss National Museum, Zurich; **Plate 8**, Aurora Borealis.

To Further Explore these Ideas and the Work of Margaret Wheatley

The ideas in *Leadership and the New Science* continue to be explored and developed by many people, not only myself. You can read about some of this exploration by visiting The Berkana Institute's website (see next page).

Please see www.margaretwheatley.com for these resources:

New writings. These are posted immediately and are downloadable for free.

Speaking Calendar. This calendar lists where and when I'm speaking in the world, and whom to contact for more information if you'd like to attend.

Books, Tapes and other Products. I have several videos, DVDs, and audio tapes available on specific topics. You can purchase these online from my site.

Seminars. If you're interested in exploring these ideas in more depth with me, please consider the different seminars I teach in several places in the world.

The Berkana Institute

The Berkana Institute was co-founded by Margaret Wheatley in 1992. Everyone engaged with Berkana, which includes people from many different countries and cultures, is experimenting with the ideas described in Leadership and the New Science.

The Berkana Institute serves people globally who are giving birth to the new forms, processes, and leadership that will restore hope to the future. Since 1992, Berkana has gradually expanded its work to reach pioneering leaders and communities in all types of organizations and in dozens of nations.

We define a leader as anyone who wants to help, who is willing to step forward to create change in their world.

The need for new leaders is urgent. We need people who can work together to resolve such pressing issues as health, poverty, hunger, illiteracy, justice, environment, democracy. We need leaders who know how to nourish and rely on the innate creativity, freedom, generosity, and caring of people. We need leaders who are life-affirming rather than life-destroying. Unless we quickly figure out how to nurture and support this new leadership, we can't hope for peaceful change. We will, instead, be confronted by increasing anarchy and societal meltdowns.

At Berkana, we know that the leaders we need are already here. Everywhere in the world, there are thousands of people stepping forward to create a future of possibility and hope. We do everything we can to support their pioneering efforts and to connect them to each other.

Initiatives of The Berkana Institute

The Berkana Exchange connects pioneering leaders who are committed to strengthening their community's leadership capacity and self-reliance by working with the wisdom and wealth already present in its people, traditions and environment. Berkana works with local Leadership Learning Centers that are focused on solving their most pressing problems—community health, ecological sustainability, economic self-reliance—by acting locally, connecting regionally and learning globally. These centers, each unique and locally designed, are in Brazil, Mexico, the U.S., Canada, Senegal, South Africa, Zimbabwe and India. More centers are joining the Exchange every year.

Berkana Learning Journeys are an opportunity to discover firsthand the new leadership emerging in the world beyond our own communities. We know the rest of the world has something essential and important to teach us about leadership. Margaret Wheatley and local pioneering leaders host groups of up to 20 people on journeys to places that challenge our view of the world, invite in disruption and open up new ways of seeing.

From the Four Directions invites people everywhere to explore their commitment to lead and to lend each other support for courageous action. We believe that leaders need to change their role from heroes to hosts. Therefore, hosting conversations is an essential leadership practice for these uncertain times. In conversation, we listen well, contemplate diverse perspectives, and are able to develop collective intelligence. In thoughtful conversation, people develop both the clarity and commitment to lead courageously. From the Four Directions develops people's capacity as hosts using a variety of conversational techniques and hosting practices.

Learn more about The Berkana Institute at www.berkana.org.

Bibliography

Abraham, Ralph. *Chaos, Gaia, Eros: A Chaos Pioneer Uncovers the Three Great Streams of History.* San Francisco: HarperSanFrancisco, 1994.

Alexander, Christopher. *The Timeless Way of Building.* New York: Oxford University Press, 1979.

Arquilla, John and David Ronfeldt. *Networks and Netwars: The Future of Terror, Crime, and Militancy.* National Defense Research Institute RAND, 2001.

Barabási, Albert-László. *LInked: The New Science of Networks.* Cambridge, MA: Perseus Publishing, 2002.

Barlow, Connie (Ed.). *From Gaia to Selfish Genes: Selected Writings in the Life Sciences.* Cambridge, MA: MIT Press, 1991.

Barnett, Thomas P.M., *Blueprint for Action.* New York: G.P. Putnam's Sons, 2005.

———. *The Pentagon's New Map.* New York: G.P. Putnam's Sons, 2004.

Bateson, Gregory. *Mind and Nature.* New York: Bantam, 1980.

Bellah, Robert N., Richard Madsen, et al. *Habits of the Heart.* New York: Harper and Row, 1985.

Blanchard, Ken, and Michael O'Connor. *Managing by Values.* San Francisco: Berrett-Koehler, 1997.

Block, Peter. *Stewardship: Choosing Service Over Self-Interest.* San Francisco: Berrett-Koehler, 1993.

Bohm, David. *Wholeness and the Implicate Order.* London: Ark Paperbacks, 1980.

Bohm, David, and Lee Nichol (Eds.). *On Dialogue.* London: Routledge, 1996.

Bok, Per. *How Nature Works: The Science of Self-Organized Criticality.* New York: Springer-Verlag, 1996.

Bonnefoy, Yves. *Mythologies.* Chicago: University of Chicago Press, 1991.

Bortoft, Henri. *The Wholeness of Nature: Goethe's Way toward a Science of Conscious Participation in Nature.* Hudson, NY: Lindisfarne, 1996.

Briggs, John, and F. David Peat. *Turbulent Mirror: An Illustrated Guide to Chaos Theory and the Science of Wholeness.* New York: Harper and Row, 1989.

Bygrave, William. "The Entrepreneurship Paradigm (I): A Philosophical Look at Its

Research Methodologies." In *Entrepreneurship Now and Then*. Baylor University, Fall 1989.

Capra, Fritjof. *The Tao of Physics*. New York: Bantam, 1976.

————. *The Turning Point: Science, Society, and the Rising Culture*. New York: Bantam, 1983.

————. *The Web of Life: A New Scientific Understanding of Living Systems*. New York: Anchor, 1996.

Capra, Fritjof, and David Steindl-Rast. *Belonging to the Universe: Explorations on the Frontiers of Science and Spirituality*. San Francisco: HarperSanFrancisco, 1991.

Cartwright, T. J. "Planning and Chaos Theory." *APA Journal* (Winter 1991): 44–56.

Chaleff, Ira. *The Courageous Follower: Standing Up To and For Our Leaders*. San Francisco: Berrett-Koehler, 1995.

Chopra, Deepak. *The New Physics of Healing*. Boulder, CO: Sounds True Recording, 1990. Audiocassette.

————. *Quantum Healing: Exploring the Frontiers of Mind and Body Science*. New York: Bantam, 1989.

Cohen, M. D., J. O. March, and J. P. Olsen. "A Garbage Can Model of Organizational Choice." *Administrative Science Quarterly*, 17 (1974): 1–25.

Cole, K. C. *Sympathetic Vibrations: Reflections on Physics as a Way of Life*. New York: Bantam, 1985.

Collins, Jim. *Good to Great: Why Some Companies Make the Leap . . . and Others Don't*. New York: HarperCollins, 2001.

Collins, James C., and Jerry I. Porras. *Built to Last: Successful Habits of Visionary Companies*. New York: HarperBusiness, 1993.

Coveney, Peter, and Roger Highfield. *The Arrow of Time: A Voyage Through Science to Solve Time's Greatest Mystery*. New York: Fawcett Columbine, 1990.

Cribbin, John. *In Search of Schroedinger's Cat: Quantum Physics and Reality*. New York: Bantam, 1984.

Crosby, Alfred W. *The Measure of Reality: Quantification and Western Society, 1250–1600*. Cambridge, UK: Cambridge University Press, 1997.

Daft, Richard, and Robert H. Lengel. *Fusion Leadership: Unlocking the Subtle Forces That Change People and Organizations*. San Francisco: Berrett-Koehler, 1998.

Davies, P. C. W., and J. Brown. *Superstrings: A Theory of Everything?* Cambridge, UK: Cambridge University Press, 1988.

DePree, Max. *Leadership Is an Art*.New York: Doubleday, 1989.

Dobbs, Betty Jo Teeter, and Margaret C. Jacob. *Newton and the Culture of Newtonianism*. Atlantic Highlands, NJ: Humanities Press, 1995.

Eglash, Ron. "Fractals in African Settlement Architecture." *Complexity*, 4.2 (Nov./Dec. 1998).

Eiseley, Loren. *The Star Thrower*. San Diego: Harvest/HBJ, 1978.

"Everything I Knew About Leadership Is Wrong: An Interview with Mort Meyerson." *Fast Company* (April-May 1996).

Feininger, Andreas. *In a Grain of Sand: Exploring Design by Nature.* San Francisco: Sierra Club Books, 1986.

Ferris, Timothy. *Coming of Age in the Milky Way.* New York: Doubleday, 1988.

Fox, Matthew. *Creation Spirituality.* San Francisco: Harper, 1991.

Frankl, Viktor. *Man's Search for Meaning.* Boston: Beacon, 1959.

Garreau, Joel. "Point Men for a Revolution: Can the Marines Survive a Shift from Hierarchies to Networks?" *Washington Post,* 6 March 1999: 1.

Gleick, James. *Chaos: Making a New Science.* New York: Viking, 1987.

Greenleaf, Robert K. *The Power of Servant-Leadership.* San Francisco: Berrett-Koehler, 1998.

Gribbin, John. *In Search of Schroedinger's Cat: Quantum Physics and Reality.* New York: Bantam, 1984.

Hamel, Gary, and C. K. Prahalad. *Competing for the Future.* Cambridge, MA: Harvard Business School Press, 1994.

Hammer, Michael. *The Reengineering Revolution.* New York: HarperBusiness, 1995.

Handy, Charles. *The Age of Unreason.* Cambridge, MA: Harvard Business School Press, 1989.

———. *Beyond Certainty: The Changing Worlds of Organizations.* Cambridge, MA: Harvard Business School Press, 1998.

———. *The Hungry Spirit: Beyond Capitalism: The Quest for Purpose in the Modern World.* New York: Broadway, 1999.

Harman, Willis, and Elisabet Sahtouris. *Biology Revisioned.* Berkeley, CA: North Atlantic Press, 1998.

Hayles, N. Katherine. *Chaos Bound: Orderly Disorder in Contemporary Literature and Science.* Ithaca: Cornell University Press, 1990.

———. *The Cosmic Web: Scientific Field Models and Literary Strategies in the Twentieth Century.* Ithaca: Cornell University Press, 1985.

Helgesen, Sally. *Web of Inclusion: A New Architecture for Building Great Organizations.* New York: Currency/Doubleday, 1995.

Heisenberg, Werner. *Physics and Philosophy.* New York: Harper Torchbooks, 1958.

Herbert, Nick. *Quantum Reality: Beyond the New Physics.* New York: Anchor Doubleday, 1985.

Hesselbein, Frances, and Paul M. Cohen (Eds.). *Leader to Leader.* New York: The Drucker Foundation, 1999.

Holman, Peggy, and Tom Devane (Eds.). *The Change Handbook: Group Methods for Shaping the Future.* San Francisco: Berrett-Koehler, 1999.

Howard, Robert. "Values Make the Company: An Interview with Robert Haas." *Harvard Business Review* (Sept.-Oct. 1990): 133–144.

Janov, Jill. *The Inventive Organization: Hope and Daring at Work.* San Francisco: Jossey-Bass, 1994.

Jantsch, Erich. *The Self-Organizing Universe.* Oxford: Pergamon, 1980.

Jaworski, Joe. *Synchronicity: The Inner Path of Leadership.* San Francisco: Berrett-Koehler, 1996.

Judson, Horace Freeland. *The Search for Solutions.* Baltimore: Johns Hopkins University Press, 1987.

Kanter, Rosabeth Moss. *The Changemasters.* New York: Simon and Schuster, 1983.

———. *Men and Women of the Corporation.* New York: Basic Books, 1977.

Kauffman, Stuart. *At Home in the Universe: The Search for Laws of Self-Organization and Complexity.* Oxford, UK: Oxford University Press, 1995.

Kelly, Kevin. *Out of Control: The Rise of Neo-Biological Civilization.* Reading, MA: Addison-Wesley, 1994.

Kuhn, Thomas. *The Structure of Scientific Revolutions.* Chicago: University of Chicago Press, 1969.

Leider, Richard J. *The Power of Purpose: Creating Meaning in Your Life and Work.* San Francisco: Berrett-Koehler, 1997.

Lessig, Lawrence. *The Future of Ideas: The Fate of the Commons in a Connected World.* New York: Vintage Books, 2002.

Lincoln, Yvonna S. (Ed.). *Organizational Theory and Inquiry: The Paradigm Revolution.* Beverly Hills, CA.: Sage, 1985.

Locke, Christopher, Rick Levine, Doc Searls, David Weinberger. *The Cluetrain Manifesto: The End of Business as Usual.* Christopher Locke, Cambridge MA: Perseus Publishing, 2001.

Louv, Richard. *Mapping the New World of Leadership.* New York: Ford Foundation, 2002.

Lovelock, J. E. *The Ages of Gaia: A Biography of our Living Earth.* New York: Norton, 1988.

———. *Gaia.* New York: Oxford University Press, 1987.

Mahr, Ernst. *Toward a New Philosophy of Biology.* Cambridge, MA: Harvard University Press, 1988.

March, Robert H. *Physics for Poets.* Chicago: Contemporary Books, 1978.

Margalef, Ramon. *Co-Evolution Quarterly* (Summer 1975): 49–66.

Margulis, Lynn. *Symbiotic Planet: A New View of Evolution.* New York: Basic Books, 1998.

Margulis, Lynn, and Dorion Sagan. *Microcosmos.* New York: Summit, 1986.

Maturana, Humberto, and Francisco Varela. *Autopoiesis and Cognition: The Realization of the Living.* London: Reidl, 1980.

McLagan, Patricia and Christo, Nel. *The Age of Participation: New Governance for the Workplace and the World.* San Francisco: Berrett-Koehler, 1995.

McLenahen, John. "Your Employees Know Better: Companies Can't Get Away with Bad Ethics Programs." *Industry Week,* 1 March 1999: 12–13.

Meadows, Donella. "Whole Earth Models and Systems." *Co-Evolution Quarterly* (Summer 1982): 98–108.

Merchant, Carolyn. *The Death of Nature: Women, Ecology and the Scientific Revolution.* New York: Harper & Row, 1980.

Mintzberg, Henry. *The Rise and Fall of Strategic Planning: Reconceiving Roles for Planning, Plans, Planners.* New York: Free Press, 1993.

Morgan, Gareth. *Images of Organization—The Executive Edition.* San Francisco: Berrett-Koehler, 1998.

"New Ideas from the Army (Really)." *Fortune,* Sept. 19, 1994, 135–139.

Nohria, N. *Creating New Business Ventures: Network Organization in Market and Corporate Contexts.* Ph.D. diss., MIT, 1988.

Nonaka, Ikujiro. "Creating Organizational Order Out of Chaos: Self-Renewal in Japanese Firms." *California Management Review* (Spring 1988): 57–73.

Nonaka, Ikujiro, and Hirotaka Takeuchi. *The Knowledge-Creating Company: How Japanese Companies Create the Dynamics of Innovation.* Oxford, UK: Oxford University Press, 1995.

Pacanowski, Michael. "Communication in the Empowering Organization." In J. A. Anderson (Ed.), *International Communications Association Yearbook II.* Beverly Hills, CA: Sage, 1988, 356–379.

Pagels, Heinz. *The Dream of Reason.* New York: Bantam, 1989.

Peat, F. David. *The Philosopher's Stone: Chaos, Synchronicity and the Hidden Order of the World.* New York: Bantam Books, 1991.

————. *Synchronicity: The Bridge Between Matter and Mind.* New York: Bantam, 1987.

Peitgen, Heinz-Otto, and Dietmar Saupe (Eds.) *The Science of Fractal Images.* New York: Springer-Verlag, 1988.

Pert, Candace, and Deepak Chopra. *The Molecules of Emotion: Why You Feel the Way You Feel.* New York: Scribner, 1997.

Petzinger, Thomas. *The New Pioneers: The Men and Women Who Are Transforming the Workplace and the Marketplace.* New York: Simon and Schuster, 1999.

Pinchot, Gifford, and Elizabeth Pinchot. *The Intelligent Organization: Engaging the Talent and Initiative of Everyone in the Workplace.* San Francisco: Berrett-Koehler, 1996.

Prahalad, C. K., and Gary Hamel. "The Core Competence of the Corporation." *Harvard Business Review* (May-June 1990): 79–91.

Prigogine, Ilya. *The End of Certainty: Time, Chaos, and the New Laws of Nature.* New York: The Free Press, 1998.

————. *Omni* (May 1983): 85–121.

Prigogine, Ilya, and Isabelle Stengers. *Order Out of Chaos.* New York: Bantam, 1984.

Rheingold, Howard. *Smart Mobs: The Next Social Revolution.* Cambridge MA: Perseus Publishing, 2002.

Rose, Steven. *Lifelines: Biology Beyond Determinism.* Oxford, UK: Oxford University Press, 1997.

Schlain, Leonard. *Art and Physics: Parallel Visions in Space, Time and Light.* New York: William Morrow, 1991.

Semler, Ricardo. "Managing Without Managers." *Harvard Business Review* (Sept.-Oct. 1989): 76–84.

Sheldrake, Rupert. *The Presence of the Past.* New York: Vintage Books, 1988.

————. *Seven Experiments That Could Change the World: A Do-It-Yourself Guide to Revolutionary Science.* New York: Riverhead, 1995.

Sheldrake, Rupert, and David Bohm. "Morphogenetic Fields and the Implicate Order." *ReVision* 5 (Fall 1982).

Smillie, Ian. *Patronage or Partnership: Local capacity building in humanitarian crises.* Bloomfield, CT: Kumarian Press and the International Development Research Centre, 2001.

Stamps, Jeffrey, and Jessica Lipnack. *The Teamnet Factor: Bringing the Power of Boundary Crossing into the Heart of Your Business.* New York: Wiley, 1995.

Starbuck, W. H. "Organizations and Their Environments." In M. D. Dunnette (Ed.) *Handbook of Industrial and Organizational Psychology.* New York: Rand, 1976, 1069–1123.

Surowiecki James. *The Wisdom of Crowds: Why the Many are Smarter Than the Few and How Collective Wisdom Shapes Business, Economies, Societies and Nations.* New York: Doubleday, 2004.

Talbot, Michael. *Beyond the Quantum.* New York: Bantam, 1986.

Tarnas, Richard. *The Passion of the Western Mind.* New York: Harmony, 1991.

The Nation. "Hurricane Gumbo" November 7, 2005.

The New York Times, December 2, 2005. "Profusion of Rebel Groups Helps Them Survive in Iraq. Dexter Filkins.

Thich Nhat Hanh. *Old Path White Clouds: Walking in the Footsteps of the Buddha.* Berkeley, CA: Parallax, 1991.

Thompson, William Irwin. *Imaginary Landscape.* New York: St. Martin's, 1989.

Thompson, William Irwin (Ed.). *Gaia 2 Emergence: The New Science of Becoming.* Hudson, NY: Lindisfarne, 1991.

Time magazine "System Failure: An investigation into what went so wrong in New Orleans." Sept 19, 2005.

Toben, Bob, and Fred Allen Wolf. *Space-Time and Beyond.* New York: Bantam, 1983.

"The Trillion Dollar Vision of Dee Hock: The Corporate Radical Who Organized Visa Wants to Dis-organize Your Company." *Fast Company,* (Oct.-Nov. 1996).

Tushman, M., and D. Nadler. "Organizing for Innovation." *California Management Review* (Spring 1986): 74–92.

USA Today, 10 March, 1999: 3.

Vaill, Peter. *Managing as a Performing Art.* San Francisco: Jossey-Bass, 1989.

Waldrop, M. Mitchell. *Complexity: The Emerging Science at the Edge of Order and Chaos.* New York: Simon and Schuster, 1992.

Weick, Karl. *The Social Psychology of Organization.* New York: Random House, 1979.

———. "Substitute for Corporate Strategy" in D. J. Teece (Ed.) *The Theoretical Context of Strategic Management.* Cambridge, MA: Ballinger, 1987.

Weinberger, David. *Small Pieces Loosely Joined: A Unified Theory of the Web.* Cambridge MA: Perseus Books, 2002.

Weisbord, Marvin. *Discovering Common Ground: How Future Search Conferences Bring People Together to Achieve Breakthrough Innovation, Empowerment, Shared Vision, Collaborative Action.* San Francisco: Berrett-Koehler, 1992.

———. *Productive Workplaces.* San Francisco: Jossey-Bass, 1987.

Weisbord, Marvin, and Sandra Janoff. *Future Search: An Action Guide to Finding Common Ground in Organizations and Communities.* San Francisco: Berrett-Koehler, 1995.

Wenger, Etienne. *Communities of Practice: Learning, Meaning, and Identity.* Cambridge, UK: Cambridge University Press, 1998.

Wheatley, Margaret J., and Myron Kellner-Rogers. *A Simpler Way.* San Francisco: Berrett-Koehler, 1996.

Whyte, David. *The Heart Aroused: Poetry and the Preservation of Soul in Corporate America.* New York: Doubleday, 1994.

Wilber, Ken. *The Holographic Paradigm and Other Paradoxes.* Boulder, CO: Shambala, 1985.

———. *Quantum Questions.* Boston: Shambala, 1984.

———. *Sex, Ecology, Spirituality: The Spirit of Evolution.* Boston: Shambala, 1995.

Wilczek, Frank, and Betsy Devine. *Longing for the Harmonies.* New York: Norton, 1988.

Willett, Carol. "Knowledge Sharing Shifts the Power Paradigm." In Mark Maybury, Daryl Morey, Bhavani Thuraisingham (Eds.). *Knowledge Management: Classic and Contemporary Works.* Cambridge, MA: Massachusetts Institute of Technology Press, 1999.

Wolf, Fred Alan. *Taking the Quantum Leap.* New York: Harper and Row, 1981.

Zohar, Danah. *The Quantum Self: Human Nature and Consciousness Defined by the New Physics.* New York: William Morrow, 1990.

Zuboff, Shoshonna. *In the Age of the Smart Machine.* New York: Basic Books, 1988.

Zukav, Gary. *The Dancing Wu Li Masters.* New York: Bantam, 1979.

Index

action-at-a-distance, confirmation of, 41–42
actions, effect on network activty, 184
Al Qaeda
    assessing strength of, 184
    bin Laden, Osama, 180, 183
    success of, 181–182
ambiguity
    control, 102
    coping skills, 101–102
Ammons, A.R., science, 156
amplification process, bifurcation point, 87–88
analysis
    information, 140
    organizational, 37–38
art, spiral forms, 81
Aspect, Alain, action-at-a-distance, 41
atomic physics, paradoxes, 5–6
attacks
    assessing terrorist network growth, 184
    defending against terrorist, 180–181
attention, shared, 149
attraction, basin of, 118
audio tapes, available Wheatley, 195
authority, jurisdiction in a disaster, 173
autonomy, and self-determination, 166–167
autopoiesis
    creation, 20
    self-reference, 85

Barabasi, Albert-Lazlo, terrorist networks,
    181–182, 185
Barnett, Thomas, the global economy "Gap", 185
Barnsley, Michael, Chaos Game, 126–128
Bateson, Gregory
    the mind, 99
    relationships, 35
behavior
    impact on organizations, 35
    individual, 142
    morphic field influence, 53
    organizational, 128–130, 149–150
    organizational fields, 54–55

beliefs, Buddhist law of dependent co-arising,
    141–142
Bell, John, action-at-a-distance, 41
Bellah, Robert N., et al., on loneliness, 32
Belousov-Zhabotinsky reaction, 81
Berkana Learning Journeys, 197
Berlin Wall, fall of, 44
bifurcation point, amplification process, 87–88
bin Laden, Osama, Al Qaeda, 180, 183
biology, theories, 12
blindness, paradigm, 174
Bohm, David
    fragmentation and wholeness, 26, 42–43
    implicate order, 43, 111
    morphic fields, 53
Bohr, Niels
    debate between Albert Einstein and, 41
    great ideas, 193
    quantum theory, 5, 32–33
books, available Wheatley, 195
boundaries
    function of, 30
    organizations, 112
    redefining relationships, 146–147
    system, 118
brain function, imagery and memory, 102–103
Briggs, John, scroll images, 81
Briggs, John and F. David Peat
    on chaos, 31
    strange attractors, 116, 118
Broccoli's fractal, 124
Buckman Laboratories
    intranet and information distribution, 104
    organizational structure, 109–110
bureaucracies, information exchange, 99
bureaucracy, risk taking, 174
Business Literacy, intelligence distribution, 110
butterfly effect, 121–122
Bygrave, William
    organizational theory, 29
    "physics envy", 160

About the Author

I have always found my attention drawn to many different disciplines: science, history, literature, systems thinking, organizational behavior, social policy, cosmology and theology. I value what I've learned from each of these different fields, because no one discipline, institution, or specialization can answer the questions that now confront us. We all must draw from many different perspectives to reweave the world.

I had an excellent liberal arts education at the University of Rochester and University College London. From 1966 to '68, I spent two years in the Peace Corps in South Korea, teaching junior high and high school English to Korean boys. On returning to the U.S., I taught junior and senior high school, then became an administrator of educational programs for children and adults who were economically poor and denied traditional educational opportunities. I received a Master of Arts degree from New York University in systems thinking and Media Ecology. My doctorate is from Harvard's program in Administration, Planning, and Social Policy, with a focus on organizational behavior and change.

I have been a consultant and speaker since 1973 and have worked, I believe, with almost all types of organizations and people. They range from the head of the U.S. Army to twelve year old Girl Scouts, from CEOs to small town ministers. This diversity includes Fortune 100 corporations, government agencies, healthcare institutions, foundations, public schools, colleges, major church denominations, professional associations, and monasteries. I have also been privileged to work on all continents. Every organization wrestles with a similar dilemma—how to maintain its integrity, direction, and effectiveness as it copes with relentless turbulence and change. But there is another similarity I'm hopeful to report: A common human desire for peace, to live together more harmoniously, more humanely.

I have served as full-time graduate management faculty at two institutions, Cambridge College in Cambridge, Massachusetts, and The Marriott School of Management, Brigham Young University, Provo, Utah. I have served in a formal advisory capacity for leadership programs in England, Croatia, Denmark. Australia and the United States, and through my work in Berkana, with leadership initiatives in India, Senegal, Zimbabwe, South Africa, Mexico, Brazil, Canada and Europe.

I am co-founder and President Emerita of The Berkana Institute, a global charitable foundation founded in 1992, and dedicated to serving life-affirming leaders. We define a leader as anyone who wants to help at this time. Berkana has worked in dozens of countries, mostly in the third world, supporting local initiatives committed to strengthening a community's leadership capacity and self-reliance by working with the wisdom and wealth already present in its people, traditions and environment. Berkana has discovered that the world is blessed with tens of thousands of courageous leaders. They are young and old, in all countries, working in all types of organizations and communities. Together, we are pioneering a new model for developing leaders who have the skills, capacity and commitment to invite their community to learn to care for itself.

For information about Berkana's work, see www.berkana.org.

Leadership and the New Science was first published in 1992, with new editions in 1999 and 2006. Each edition contains new material on where the ideas of new science are evident in the world. This book is credited with establishing a fundamentally new approach to how we think about organizations. It has been translated into seventeen languages and won many awards, including "Best Management Book of 1992" in Industry Week, Top Ten Business Books of the 1990s by CIO Magazine, and Top Ten Business books of all time by Xerox Corporation. The video of *Leadership and the New Science,* produced by CRM films, has also won several film awards.

In 1996, I co-authored *A Simpler Way* (with Myron Kellner-Rogers). *A Simpler Way* explores the question: Could we organize human endeavor differently if we understood how Life organizes? Through photos, poetry, and prose, the book contemplates self-organization, and the conditions that nurture life and organizations.

Turning to One Another: Simple Conversations to Restore Hope to the Future (2002) is written from the belief that we could change the world if we just begin listening to one another again. Great social change movements always begin from the simple act of friends talking to each other about their fears and dreams. This book contains ten conversation starters as well as guidance about how to host good conversations.

Finding Our Way: Leadership for an Uncertain Time (2005) is a collection of my practice-focused articles, where I apply themes addressed throughout my career to detail the organizational practices and behaviors that bring them to life. These pieces represent more than a decade of work, of how I took the ideas in my earlier books and applied them in practice in many different situations. However, this is more than a collection of articles. I updated, revised or substantially added to the original content of each one. In this way, everything written here represents my most current views on these subjects.

My articles appear in a wide range of professional publications and magazines, and can be downloaded free from my website.

www.margaretwheatley.com

On this website, you will also see a number of videos, CDs and DVDs that I've produced on different organizational and leadership topics.

I was raised on the East Coast of the U.S., first in the New York City area, and then in Boston. In 1989, my family and I moved west to the mountains and red rocks of Utah. I have two adult sons, five stepchildren, and fourteen grandchildren. My family, friends and work bring me reliable joy, and so does the time I spend with my horses, or hiking and skiing in the true quiet of wilderness.

I can be reached at:

Margaret J. Wheatley Inc.
P.O. Box 1407
Provo, Utah 84603
Tel: 801-377-2996
Fax: 801-377-2998
www.margaretwheatley.com

About Berrett-Koehler Publishers

Berrett-Koehler is an independent publisher dedicated to an ambitious mission: Creating a World that Works for All.

We believe that to truly create a better world, action is needed at all levels — individual, organizational, and societal. At the individual level, our publications help people align their lives with their values and with their aspirations for a better world. At the organizational level, our publications promote progressive leadership and management practices, socially responsible approaches to business, and humane and effective organizations. At the societal level, our publications advance social and economic justice, shared prosperity, sustainability, and new solutions to national and global issues.

A major theme of our publications is "Opening Up New Space." They challenge conventional thinking, introduce new ideas, and foster positive change. Their common quest is changing the underlying beliefs, mindsets, institutions, and structures that keep generating the same cycles of problems, no matter who our leaders are or what improvement programs we adopt.

We strive to practice what we preach — to operate our publishing company in line with the ideas in our books. At the core of our approach is *stewardship*, which we define as a deep sense of responsibility to administer the company for the benefit of all of our "stakeholder" groups: authors, customers, employees, investors, service providers, and the communities and environment around us.

We are grateful to the thousands of readers, authors, and other friends of the company who consider themselves to be part of the "BK Community." We hope that you, too, will join us in our mission.

Be Connected

Visit Our Website

Go to www.bkconnection.com to read exclusive previews and excerpts of new books, find detailed information on all Berrett-Koehler titles and authors, browse subject-area libraries of books, and get special discounts.

Subscribe to Our Free E-Newsletter

Be the first to hear about new publications, special discount offers, exclusive articles, news about bestsellers, and more! Get on the list for our free e-newsletter by going to www.bkconnection.com.
[A head]Participate in the Discussion
To see what others are saying about our books and post your own thoughts, check out our blogs at www.bkblogs.com.

Get Quantity Discounts

Berrett-Koehler books are available at quantity discounts for orders of ten or more copies. Please call us toll-free at (800) 929-2929 or email us at bkp.orders@aidcvt.com.

Host a Reading Group

For tips on how to form and carry on a book reading group in your workplace or community, see our website at www.bkconnection.com.

Join the BK Community

Thousands of readers of our books have become part of the "BK Community" by participating in events featuring our authors, reviewing draft manuscripts of forthcoming books, spreading the word about their favorite books, and supporting our publishing program in other ways. If you would like to join the BK Community, please contact us at bkcommunity@bkpub.com.